Selected Titles in This Series

687 **Guy David and Stephen Semmes,** Uniform rectifiability and quasiminimizing sets of arbitrary codimension, 2000

686 **L. Gaunce Lewis, Jr.,** Splitting theorems for certain equivariant spectra, 2000

685 **Jean-Luc Joly, Guy Metivier, and Jeffrey Rauch,** Caustics for dissipative semilinear oscillations, 2000

684 **Harvey I. Blau, Bangteng Xu, Z. Arad, E. Fisman, V. Miloslavsky, and M. Muzychuk,** Homogeneous integral table algebras of degree three: A trilogy, 2000

683 **Serge Bouc,** Non-additive exact functors and tensor induction for Mackey functors, 2000

682 **Martin Majewski,** Rational homotopical models and uniqueness, 2000

681 **David P. Blecher, Paul S. Muhly, and Vern I. Paulsen,** Categories of operator modules (Morita equivalence and projective modules), 2000

680 **Joachim Zacharias,** Continuous tensor products and Arveson's spectral C^*-algebras, 2000

679 **Y. A. Abramovich and A. K. Kitover,** Inverses of disjointness preserving operators, 2000

678 **Wilhelm Stannat,** The theory of generalized Dirichlet forms and its applications in analysis and stochastics, 1999

677 **Volodymyr V. Lyubashenko,** Squared Hopf algebras, 1999

676 **S. Strelitz,** Asymptotics for solutions of linear differential equations having turning points with applications, 1999

675 **Michael B. Marcus and Jay Rosen,** Renormalized self-intersection local times and Wick power chaos processes, 1999

674 **R. Lawther and D. M. Testerman,** A_1 subgroups of exceptional algebraic groups, 1999

673 **John Lott,** Diffeomorphisms and noncommutative analytic torsion, 1999

672 **Yael Karshon,** Periodic Hamiltonian flows on four dimensional manifolds, 1999

671 **Andrzej Rosłanowski and Saharon Shelah,** Norms on possibilities I: Forcing with trees and creatures, 1999

670 **Steve Jackson,** A computation of δ^1_5, 1999

669 **Seán Keel and James McKernan,** Rational curves on quasi-projective surfaces, 1999

668 **E. N. Dancer and P. Poláčik,** Realization of vector fields and dynamics of spatially homogeneous parabolic equations, 1999

667 **Ethan Akin,** Simplicial dynamical systems, 1999

666 **Mark Hovey and Neil P. Strickland,** Morava K-theories and localisation, 1999

665 **George Lawrence Ashline,** The defect relation of meromorphic maps on parabolic manifolds, 1999

664 **Xia Chen,** Limit theorems for functionals of ergodic Markov chains with general state space, 1999

663 **Ola Bratteli and Palle E. T. Jorgensen,** Iterated function systems and permutation representation of the Cuntz algebra, 1999

662 **B. H. Bowditch,** Treelike structures arising from continua and convergence groups, 1999

661 **J. P. C. Greenlees,** Rational S^1-equivariant stable homotopy theory, 1999

660 **Dale E. Alspach,** Tensor products and independent sums of $\mathcal{L}_p$-spaces, $1 < p < \infty$, 1999

659 **R. D. Nussbaum and S. M. Verduyn Lunel,** Generalizations of the Perron-Frobenius theorem for nonlinear maps, 1999

658 **Hasna Riahi,** Study of the critical points at infinity arising from the failure of the Palais-Smale condition for n-body type problems, 1999

657 **Richard F. Bass and Krzysztof Burdzy,** Cutting Brownian paths, 1999

656 **W. G. Bade, H. G. Dales, and Z. A. Lykova,** Algebraic and strong splittings of extensions of Banach algebras, 1999

(Continued in the back of this publication)

Uniform Rectifiability and Quasiminimizing Sets of Arbitrary Codimension

Memoirs
of the
American Mathematical Society

Number 687

Uniform Rectifiability and
Quasiminimizing Sets
of Arbitrary Codimension

Guy David
Stephen Semmes

March 2000 • Volume 144 • Number 687 (end of volume) • ISSN 0065-9266

American Mathematical Society
Providence, Rhode Island

2000 *Mathematics Subject Classification.* Primary 49Q20; Secondary 28A75, 42B99.

Library of Congress Cataloging-in-Publication Data

David, Guy, 1957–
  Uniform rectifiability and quasiminimizing sets of arbitrary codimension / Guy David, Stephen Semmes.
    p. cm. — (Memoirs of the American Mathematical Society, ISSN 0065-9266 ; no. 687)
  "March 2000, volume 144, number 687 (end of volume)."
  Includes bibliographical references.
  ISBN 0-8218-2048-6 (alk. paper)
  1. Minimal surfaces. 2. Geometric measure theory. 3. Fourier analysis. I. Semmes, Stephen, 1962– II. Title. III. Series.
QA3.A57 no. 687
[QA644]
510 s—dc21
[516.3′62]   99-058340

Memoirs of the American Mathematical Society

This journal is devoted entirely to research in pure and applied mathematics.

Subscription information. The 2000 subscription begins with volume 143 and consists of six mailings, each containing one or more numbers. Subscription prices for 2000 are $466 list, $419 institutional member. A late charge of 10% of the subscription price will be imposed on orders received from nonmembers after January 1 of the subscription year. Subscribers outside the United States and India must pay a postage surcharge of $30; subscribers in India must pay a postage surcharge of $43. Expedited delivery to destinations in North America $35; elsewhere $130. Each number may be ordered separately; *please specify number* when ordering an individual number. For prices and titles of recently released numbers, see the New Publications sections of the *Notices of the American Mathematical Society*.

Back number information. For back issues see the *AMS Catalog of Publications*.

Subscriptions and orders should be addressed to the American Mathematical Society, P. O. Box 5904, Boston, MA 02206-5904. *All orders must be accompanied by payment.* Other correspondence should be addressed to Box 6248, Providence, RI 02940-6248.

Copying and reprinting. Individual readers of this publication, and nonprofit libraries acting for them, are permitted to make fair use of the material, such as to copy a chapter for use in teaching or research. Permission is granted to quote brief passages from this publication in reviews, provided the customary acknowledgment of the source is given.

Republication, systematic copying, or multiple reproduction of any material in this publication is permitted only under license from the American Mathematical Society. Requests for such permission should be addressed to the Assistant to the Publisher, American Mathematical Society, P. O. Box 6248, Providence, Rhode Island 02940-6248. Requests can also be made by e-mail to reprint-permission@ams.org.

Memoirs of the American Mathematical Society is published bimonthly (each volume consisting usually of more than one number) by the American Mathematical Society at 201 Charles Street, Providence, RI 02904-2294. Periodicals postage paid at Providence, RI. Postmaster: Send address changes to Memoirs, American Mathematical Society, P. O. Box 6248, Providence, RI 02940-6248.

© 2000 by the American Mathematical Society. All rights reserved.
This publication is indexed in *Science Citation Index*®, *SciSearch*®, *Research Alert*®, *CompuMath Citation Index*®, *Current Contents*®/*Physical, Chemical & Earth Sciences*.
Printed in the United States of America.

∞ The paper used in this book is acid-free and falls within the guidelines established to ensure permanence and durability.
Visit the AMS home page at URL: http://www.ams.org/

10 9 8 7 6 5 4 3 2 1   05 04 03 02 01 00

Contents

Chapter 0.	Introduction	1
Chapter 1.	Quasiminimizers	7
Chapter 2.	Uniform Rectifiability and the Main Result	10
Chapter 3.	Lipschitz Projections into Skeleta	13
Chapter 4.	Local Ahlfors-Regularity	20
Chapter 5.	Lipschitz Mappings with Big Images	27
Chapter 6.	From Lipschitz Functions to Projections	30
Chapter 7.	Regular Sets and Cubical Patchworks	33
Chapter 8.	A Stopping-Time Argument	44
Chapter 9.	Proof of Main Lemma 8.7	51
9.1.	A general deformation result	51
9.2.	Application to quasiminimizers	67
Chapter 10.	Big Projections	78
Chapter 11.	Restricted and Dyadic Quasiminimizers	85
Chapter 12.	Applications	97
12.1.	The initial set-up	98
12.2.	Stability of sets	104
12.3.	Topological interpretations	112
12.4.	Polyhedral approximations and minimizers	114
12.5.	General sets	120
Bibliography		131

ABSTRACT. Roughly speaking, a d-dimensional subset of $\mathbf{R}^n$ is *minimizing* if arbitrary deformations of it (in a suitable class) cannot decrease its d-dimensional volume. For *quasiminimizing* sets, one allows the mass to decrease, but only in a controlled manner. To make this precise we follow Almgren's notion of "restricted sets" [**2**]. Graphs of Lipschitz mappings $f : \mathbf{R}^d \to \mathbf{R}^{n-d}$ are always quasiminimizing, and Almgren showed that quasiminimizing sets are *rectifiable*. Here we establish uniform rectifiability properties of quasiminimizing sets, which provide a more quantitative sense in which these sets behave like Lipschitz graphs. (Almgren also established stronger smoothness properties under tighter quasiminimality conditions.)

Quasiminimizing sets can arise as minima of functionals with highly irregular "coefficients". For such functionals, one cannot hope in general to have much more in the way of smoothness or structure than uniform rectifiability, for reasons of bilipschitz invariance. (See also [**9**].)

One motivation for considering minimizers of functionals with irregular coefficients comes from the following type of question. Suppose that one is given a compact set K with upper bounds on its d-dimensional Hausdorff measure, and lower bounds on its d-dimensional topology. What can one say about the structure of K? To what extent does it behave like a nice d-dimensional surface? A basic strategy for dealing with this issue is to first replace K by a set which is minimizing for a measurement of volume that imposes a large penalty on points which lie outside of K. This leads to a kind of regularization of K, in which cusps and very scattered parts of K are removed, but without adding more than a small amount from the complement of K. The results for quasiminimizing sets then lead to uniform rectifiability properties of this regularization of K.

To actually produce minimizers of general functionals it is sometimes convenient to work with (finite) discrete models. A nice feature of uniform rectifiability is that it provides a way to have bounds that cooperate robustly with discrete approximations, and which survive in the limit as the discretization becomes finer and finer.

1991 *Mathematics Subject Classification*. Primary 49Q20; Secondary 28A75, 42B99.

Key words and phrases. Quasiminimal surfaces, Almgren restricted sets, uniform rectifiability, topological nondegeneracy conditions, polyhedral approximations, dyadic minimal surfaces.

CHAPTER 0

Introduction

Roughly speaking, a *quasiminimizing set* is one whose mass cannot be decreased too much through a modest deformation. To make this precise, let us first recall a few basic definitions.

Fix integers d and n with $0 < d < n$. If E is a subset of $\mathbf{R}^n$, then we define the *d-dimensional Hausdorff measure* $H^d(E)$ of E as follows. Given $\delta > 0$, we first set

$$(0.1) \quad H^d_\delta(E) = \inf \Big\{ \sum_j (\operatorname{diam} A_j)^d : \{A_j\} \text{ is a sequence of sets in } \mathbf{R}^n$$
$$\text{which covers } E \text{ and satisfies}$$
$$\operatorname{diam} A_j < \delta \text{ for all } j \Big\}.$$

It is easy to see that $H^d_\delta(E)$ can only become larger as δ gets smaller. Thus the limit

$$(0.2) \quad \lim_{\delta \to 0} H^d_\delta(E)$$

exists (but may be infinite), and $H^d(E)$ is defined to be the value of this limit. (See [**12, 13, 20**] for more information.)

A mapping $f : \mathbf{R}^n \to \mathbf{R}^n$ is said to be *Lipschitz* if there is a constant $C > 0$ so that

$$(0.3) \quad |f(x) - f(y)| \leq C\,|x - y|$$

for all $x, y \in \mathbf{R}^n$. Sometimes we might say that f is C-*Lipschitz* to be more precise. If f is C-Lipschitz, then

$$(0.4) \quad H^d(f(A)) \leq C^d\,H^d(A)$$

for all subsets A of $\mathbf{R}^n$. This is not hard to check, just using the definitions.

Let S be a closed and unbounded subset of $\mathbf{R}^n$ such that $H^d(S \cap B)$ is finite for every ball $B \subseteq \mathbf{R}^n$. Fix a number $k \geq 1$. We say that S is k-*quasiminimizing* if

$$(0.5) \quad H^d(S \cap W) \leq k\,H^d(\phi(S \cap W))$$

whenever $\phi : \mathbf{R}^n \to \mathbf{R}^n$ is a Lipschitz mapping such that

$$(0.6) \quad W = \{x \in \mathbf{R}^n : \phi(x) \neq x\}$$

is bounded. We call S a quasiminimizing set if it is k-quasiminimizing for some k.

When $k = 1$ this is a version of the familiar "volume-minimizing" property, as in [**3, 13, 21, 27**] (although one should be a bit careful about the issue of multiplicities). In the present formulation we are following Almgren [**2**], who used the name "restricted sets".

Received by the editor July 28, 1998.

The first author was supported by the Institut Universitaire de France, and the second author was partially supported by the U.S. National Science Foundation.

If S is a d-plane, then it is not too difficult to verify that (0.5) holds with $k = 1$. To obtain some quasiminimizing sets with larger values of k one can employ the following method. Let g be a one-to-one mapping from $\mathbf{R}^n$ onto itself such that g and g^{-1} are both C-Lipschitz. If P is a d-plane in $\mathbf{R}^n$, then $g(P)$ is k-quasiminimizing with $k = C^{2d}$. This is not hard to check. (See also p54-5 of [**2**].)

As a special case of this, imagine that $\mathbf{R}^n$ is decomposed as the orthogonal direct sum of a d-plane P and an $(n-d)$-plane Q (both passing through the origin, say), and let $h : P \to Q$ be any Lipschitz mapping. If $g : \mathbf{R}^n \to \mathbf{R}^n$ is defined by setting

$$(0.7) \qquad g(p+q) = p + q + h(p)$$

for $p \in P$ and $q \in Q$, then g and its inverse are both Lipschitz mappings, and the preceding observation implies that

$$(0.8) \qquad g(P) = \{p + h(p) : p \in P\}$$

is a quasiminimizing set. In other words, graphs of Lipschitz mappings from P to Q are always quasiminimizing.

This shows that quasiminimizing sets need not be smooth in general, no more smooth than a Lipschitz graph. By contrast, sets which are actually minimizing (1-quasiminimizing) are necessarily smooth (and even real-analytic) almost everywhere, as in [**1**]. If a set is approximately minimal at small scales, in the sense that the constant k in (0.5) approaches 1 in a suitably-controlled manner when the displacement ϕ is sufficiently localized, then there are results which give $C^{1,\alpha}$-smoothness almost everywhere due to Almgren [**2**]. (In fact, Almgren allows almost-minimality with respect to a wider class of measurements, and not just volume.)

For arbitrary quasiminimizers S, Almgren [**2**] establishes classical rectifiability properties which imply that S behaves like a C^1-smooth set except on a set of small measure. This is consistent with the examples of Lipschitz graphs and bilipschitz images of d-planes, because Lipschitz mappings between Euclidean spaces can be modified on sets of small measure to produce C^1 mappings, as is well known. (See [**13**].) However, this kind of statement is qualitative rather than quantitative. For a Lipschitz graph, for instance, it says nothing about the Lipschitz norm of the function being graphed, or anything like that.

Here we seek uniform bounds for the geometric structure of quasiminimizers. Specifically, we shall show that a quasiminimizing set S can always be realized as a union of a set of H^d-measure 0 and a set S^* which is "Ahlfors regular" and "uniformly rectifiable". The precise definitions of these properties will be given in Chapter 2. Roughly speaking, "Ahlfors regularity" means upper and lower bounds for the mass of a set inside a given ball, bounds which are like those for d-planes and their bilipschitz images. This was also obtained by Almgren before, and it is the uniform rectifiability which is really the main point now. We shall establish a slightly stronger condition, which is that S^* contains "big pieces of Lipschitz graphs". (See Chapter 2 for the definition.)

One of the reasons for looking at quasiminimizers in general is that they include minimizers for many functionals, including functionals with highly nonsmooth coefficients. For instance, instead of measuring volume in the usual way, one can employ measurements of the form

$$(0.9) \qquad J_M(S) = H^d(S \cap F) + M \cdot H^d(S \backslash F),$$

where F is a fixed closed set and M is a positive real number. When M is large, this has the effect of putting a high cost on parts of S which lie outside of F.

The minimization of functionals like (0.9) (with suitable boundary or topological conditions on S) can be a useful tool for regularizing a given set F. Namely, it can have the effect of replacing F with a set which is nearly contained in F, but for which cusps or very scattered parts (or other types of singularities) are removed. In this we are motivated by an argument of Morel and Solimini, for "regularizing" a rectifiable curve that contains a given set in a simple way through minimization.

As a general scenario, imagine that our set F is compact, has d-dimensional measure bounded from above by some constant C_0, and enjoys nontrivial lower bounds for its d-dimensional topological behavior. For example, F might satisfy a (quantitative) linking condition with a set of dimension $n-d-1$ (inside $\mathbf{R}^n$), or there might be a homotopically-nontrivial mapping f from F to the unit d-sphere $\mathbf{S}^d$, with a bound on the Lipschitz norm of f. In circumstances like these, it is natural to hope that F contains a substantial subset of a d-dimensional Lipschitz graph, with bounds for the Lipschitz constant for the graph in terms of the bounds in the assumptions on F. Otherwise, one might expect that F would be too scattered to support nontrivial d-dimensional topological behavior in a quantitative way.

In the context of ordinary rectifiability, the "scatteredness" of unrectifiable sets can often be analyzed through Federer's structure theorem [13, 20], which characterizes rectifiability and unrectifiability of d-dimensional sets in terms of measures of projections of sets onto d-planes. Such a tool is (so far) not available for quantitative control on geometric complexity, like uniform rectifiability or big pieces of Lipschitz graphs.

We shall consider this type of question (about the behavior of sets with upper bounds on d-dimensional Hausdorff measure and lower bounds for d-dimensional topology) in some detail Chapter 12. The following statement contains some of the basic information which will be derived there.

THEOREM 0.10. *Fix an integer j_0, and let A be a subset of $\mathbf{R}^n$ which is a finite union of dyadic cubes of sidelength 2^{-j_0} in $\mathbf{R}^n$. (In other words, A is a finite union of cubes in Δ_{j_0}, where Δ_{j_0} is defined just before Proposition 11.13.) Let $\theta > 0$ be as in Proposition 12.46 (in Section 12.2), let E be a compact subset of A with $H^d(E) < \infty$, and assume that E satisfies the "stability" property (12.47) in Proposition 12.46. (Roughly speaking, this is a quantitative condition concerning d-dimensional topological nondegeneracy of E. It says that E cannot be collapsed down to a set of H^d-measure less than θ by a continuous mapping from E into A which is homotopic to the identity through mappings from E into A. With the choice of θ in Proposition 12.46 (which depends only on j_0 and n), the alternative is that E can be collapsed down to a set of dimension $d-1$ in A by a mapping which is homotopic to the identity through mappings from E into A.)*

Under these conditions, given any $\tau > 0$, there is a compact set $Z \subseteq \mathbf{R}^n$ such that

$$(0.11) \qquad Z \text{ is Ahlfors regular, uniformly rectifiable, and}$$
$$\text{contains big pieces of Lipschitz graphs,}$$

(0.12) $$H^d(Z) \geq \theta',$$

and

(0.13) $$H^d(Z \backslash E) \leq \tau H^d(E).$$

The constant θ' in (0.12) depends only on n and j_0 (and not on τ in particular), while the constants for the conditions in (0.11) are bounded in a way that depends only on n, j_0, $\operatorname{diam} A$, and τ.

The constant θ' comes up in a way that is very similar to θ, in Proposition 12.61 in place of Proposition 12.46. Both θ and θ' can be taken to be in the form of a constant that depends only on n times $2^{-j_0 d}$.

Theorem 0.10 follows from the combination of Proposition 12.116 and Theorems 12.122 and 12.125, all in Section 12.5.

REMARKS 0.14. (a) The stability condition in the hypothesis of Theorem 0.10 is implied by suitable (quantitative) linking conditions, or conditions of homotopic-nontriviality of maps into $\mathbf{S}^d$, of the type mentioned shortly before the statement of Theorem 0.10. This is discussed in Section 12.3. For that matter, all reasonable conditions of this general nature are closely connected to each other, because E is required to have finite H^d-measure (and hence topological dimension $\leq d$) anyway. See [**16**] for related information. The stability condition makes a good "base" from which to work, and Proposition 12.46 provides a clear dichotomy between it and a natural "instability" property.

(b) The formulation of Theorem 0.10 contains nontrivial information only when τ is small enough so that $H^d(Z \cap E)$ is forced to be positive by (0.12) and (0.13). Otherwise, one could choose Z in a manner which has nothing to do with E. The actual process by which Z is obtained from E in Section 12.5 makes a stronger connection between the two sets than is indicated in the statement of Theorem 0.10, and we shall return to this in (d) below.

(c) Roughly speaking, one can think of the set Z in Theorem 0.10 as being obtained by minimizing a functional like (0.9). This is not quite what we do here, and indeed one has to be somewhat careful about existence of minimizers. Existence issues are often treated through the use of currents or varifolds (as in [**13, 27**]), but here we take a more primitive approach. Namely, we look at minimizers of functionals like (0.9) among *finite* collections of polyhedral sets. In this way, the existence of minimizers becomes trivial, but we have to be careful to get uniform estimates which do not depend on the class of competitors. That is, the class of competitors would normally be something like all (dyadic) d-dimensional polyhedra in a fixed region and at a fixed degree of resolution, and it will be important to have estimates that do not blow up as the degree of resolution goes to 0. We shall do this by showing that minimizers over suitable finite collections of polyhedral sets are also quasiminimizers in the general sense (without the restriction to polyhedral variations), and with uniform bounds on the constants.

This approach reflects some nice features of the notions of uniform rectifiability and "big pieces of Lipschitz graphs"; they are sufficiently robust to cooperate with polyhedral approximation in a simple way, and also strong enough so that uniform bounds for (finite) polyhedral sets contain nontrivial information that survives in limiting regimes for the scale of discretization. In the context of Theorem 0.10, the set Z will in fact be derived from sequences of minimizers for finite problems using Hausdorff limits. (More precisely, Z will not be taken as a limit of actual minimizing sets, but instead we shall have to first "prune" the minimizers by removing some collapsed pieces of H^d-measure 0.)

(d) The lower bound (0.12) in Theorem 0.10 provides a kind of nondegeneracy for Z, but in fact the construction in Chapter 12 gives much more precise information. Specifically, the polyhedral approximations to Z mentioned in (c) have approximately the same kind of topological information as the original set E; this follows from the definitions employed in Sections 12.4 and 12.5 (i.e., for the classes of (essentially) polyhedral sets used as competitors for the minimization of functionals like (0.9)). To some extent the topological information survives in the passage to the limit that produces Z, but it is not quite as simple as for the finite approximations. This is discussed further in Remark 12.171 at the end of Section 12.5.

(e) For the dependence on the parameters j_0, diam A, and $H^d(E)$ in Theorem 0.10, one should keep in mind that the whole story is scale-invariant. One could just as well restrict oneself to $j_0 = 0$, for instance. The diameter of A does not play much of an important role in any case. That is, the bounds for Ahlfors regularity, uniform rectifiability, and big pieces of Lipschitz graphs do not depend on the diameter of A when one restricts oneself to scales no greater than 2^{-j_0}; the dependence on diam A comes from the (brutal) extrapolation of these bounds to scales between 2^{-j_0} and diam A. (The basic method does not really say anything about what happens at scales larger than 2^{-j_0} anyway.)

(f) Codimension-1 versions of this problem were treated earlier in [**17**] and [**9, 10**], and by different methods (from each other). The approach of [**9, 10**] is similar to the one given here, except that it is based on simpler notions of quasiminimizers. In particular, one can make use of "functions of bounded variation" (as in [**14**]) in a convenient way for the codimension-1 case.

See [**25**] for further discussion of the relationship between topological and measure-theoretic properties of sets.

The definition of quasiminimizers that we shall actually use in the main text is somewhat more complicated than the one given above, and it allows for localized forms of quasiminimality. This will be discussed in detail in Chapter 1. The precise statement of the main result about quasiminimizing sets is given in Chapter 2, along with the relevant definitions, and the proof of it is given in Chapters 3 - 10. The main engine of the argument is contained in Chapters 5, 8, and 9, while Chapters 3, 6, and 7, although clearly needed, are of a more technical and predictable nature. Chapter 4 provides a proof of Almgren's local Ahlfors-regularity result, and Chapter 10 establishes the property of "big pieces of Lipschitz graphs" for quasiminimizers once uniform rectifiability is known. In Chapter 11 we give some tools for dealing with discretized versions of quasiminimizing sets, e.g., for showing that discrete quasiminimizers are ordinary quasiminimizers (with slightly worse constants) under suitable conditions. As in Remark 0.14 (c), part of the point here is that it is very easy to obtain minimizers (and hence quasiminimizers) in discrete settings, and one wants to be able to have uniform bounds for their geometric behavior. Chapter 12 concerns applications of the results on quasiminimizers to analysis of sets with upper bounds for H^d-measure and lower bounds for d-dimensional topology, as indicated above.

Portions of this work were performed during visits of the second author to the Institut des Hautes Études Scientifiques in Bures-sur-Yvette, France, to which the authors are grateful. The authors would also like to thank William Beckner for his efforts as editor, and the referees for their helpful comments and suggestions.

CHAPTER 1

Quasiminimizers

Let U be an open set in $\mathbf{R}^n$, and fix an integer d, $0 < d < n$, and constants $k \in [1, +\infty)$ and $\delta \in (0, +\infty]$. Let S be a subset of U which is nonempty and relatively closed, i.e.,

(1.1) $$S \neq \phi \quad \text{and} \quad \overline{S}\backslash S \subseteq \mathbf{R}^n \backslash U.$$

Assume also that

(1.2) $$H^d(S \cap B) < +\infty \quad \text{for each ball } B \subseteq U,$$

where H^d denotes d-dimensional Hausdorff measure. To say that S is a quasiminimizer for H^d means that a certain condition of comparison with deformations of S is satisfied. Before we state this precisely, we need some auxiliary definitions and notation.

Our deformations of S will be sets of the form $\phi(S)$, where

(1.3) $$\phi : \mathbf{R}^n \to \mathbf{R}^n \text{ is Lipschitz}$$

and has the following properties. Set

(1.4) $$W = \{x \in \mathbf{R}^n : \phi(x) \neq x\}.$$

We require that

(1.5) $$\operatorname{diam}(W \cup \phi(W)) < \delta$$

and

(1.6) $$\operatorname{dist}(W \cup \phi(W), \mathbf{R}^n \backslash U) > 0.$$

We also ask that ϕ be homotopic to the identity through Lipschitz mappings that satisfy (1.5) and (1.6), as in the following condition:

(1.7) there is a continuous mapping $h : [0, 1] \times \mathbf{R}^n \to \mathbf{R}^n$ such that

(1.7.a) $h(0, x) = x$ and $h(1, x) = \phi(x)$ for all $x \in \mathbf{R}^n$,

(1.7.b) $h(t, \cdot)$ is Lipschitz for each $t \in [0, 1]$,

and

(1.7.c) $\operatorname{diam} \widehat{W} < \delta$ and $\operatorname{dist}(\widehat{W}, \mathbf{R}^n \backslash U) > 0.$

Here $\widehat{W}$ is the set of points in $\mathbf{R}^n$ which are of the form $h(t, x)$ or x, where $x \in \mathbf{R}^n$ and $t \in [0, 1]$ satisfy $h(t, x) \neq x$. In other words, $\widehat{W}$ is the union of the sets $W_t \cup \phi_t(W_t)$, $0 \leq t \leq 1$, where W_t is associated to $\phi_t(x) = h(t, x)$ as in (1.4).

The precise form of our comparison condition is

(1.8) $$H^d(S \cap W) \leq k H^d\Big(\phi(S \cap W)\Big) \text{ for all Lipschitz mappings}$$
ϕ which satisfy (1.5), (1.6), and (1.7).

Let us summarize the definition of a quasiminimizer as follows.

DEFINITION 1.9. *Let $0 < d < n$, an open set $U \subseteq \mathbf{R}^n$, and constants $k \in [1, +\infty)$ and $\delta \in (0, +\infty]$ be fixed. A (U, k, δ)-quasiminimizer for H^d (or "quasiminimizer", for short) is a subset S of U which satisfies (1.1), (1.2), and (1.8).*

This is a minor variation of what Almgren calls a (k, δ)-restricted set in [2]. The differences are that Almgren also requires that $\operatorname{diam} S < +\infty$ and $H^d(S) < +\infty$, and he does not ask the competitors ϕ to satisfy the homotopy condition (1.7). We shall say more about the latter in a moment.

If $U = \mathbf{R}^n$ and $\delta = +\infty$, then the above definition of a quasiminimizer simplifies to the one given in the introduction. That is, a nonempty closed set S is a quasiminimizer in that case if it has locally finite Hausdorff measure and satisfies

$$H^d(S \cap W) \leq k H^d\big(\phi(S \cap W)\big)$$

for every Lipschitz mapping $\phi : \mathbf{R}^n \to \mathbf{R}^n$ such that $W = \{x \in \mathbf{R}^n : \phi(x) \neq x\}$ is bounded. It is easy to see that S is necessarily unbounded in this situation.

Note that if the set W in (1.4) is bounded, then $\phi(\mathbf{R}^n) = \mathbf{R}^n$ automatically holds by standard results in topology, and therefore $\phi(W) \supseteq W$. Thus we could replace $W \cup \phi(W)$ with $\phi(W)$ in (1.5) and (1.6) without losing any information.

Although we require that ϕ be Lipschitz in (1.3), it is important that we do not impose any bound on the Lipschitz constant C of ϕ. Similarly, we do not require specific bounds on the homotopy h in (1.7). In many of our arguments, ϕ will have an extremely large Lipschitz constant, over which we have no control.

REMARK 1.10. If the set $W \cup \phi(W)$ is constrained in a ball or a cube B such that $\operatorname{diam} B < \delta$ and $\operatorname{dist}(B, \mathbf{R}^n \backslash U) > 0$, then (1.7) holds automatically. This is an easy consequence of the convexity of B; one can simply take $h(t, x) = t\phi(x) + (1-t)x$.

The quasiminimizing condition (1.8) is a bit technical, but it is also quite convenient, for the way that it accommodates many different types of situations. Notice that a natural tendency for a quasiminimizer is simply to collapse to a set of (d-dimensional) measure 0. In practice this is often prevented by the restrictions (1.5), (1.6), and (1.7) on the mapping ϕ. For instance, (1.6) can have the effect of fixing "boundary conditions" which prevent collapse. This is illustrated by the example of the intersection $S = P \cap B$ of a d-plane P with an open ball B that meets P. This is a $(B, 1, +\infty)$-quasiminimizer, as one can verify. As a different type of example, imagine that we take U to be the complement of a ball in $\mathbf{R}^n$, and that S is a compact subset of U which surrounds the ball, i.e., separates it from infinity. This separation again prevents collapsing (with $d = n - 1$). In higher codimensions one can have more complicated linking conditions, and we shall pursue related themes in Chapter 12, especially Sections 12.2 and 12.3.

One can also take advantage of the parameter δ to obtain nondegenerate quasiminimizers. For example, given any $k > 1$, the unit sphere in $\mathbf{R}^n$ will be an $(\mathbf{R}^n, k, \delta)$-quasiminimizer as soon as δ is small enough (depending on k).

REMARK 1.11. If S is a (U, k, δ)-quasiminimizer and V is an open subset of U such that $S \cap V \neq \phi$, then $S \cap V$ is a (V, k, δ)-quasiminimizer with the same values of k and δ. This is a trivial consequence of the definitions, but it is useful to remember. In fact, our main results can all be derived from the special case in which V is an open ball.

Given a d-dimensional quasiminimizing set S in an open set U, define a new set S^* by

$$\begin{aligned}(1.12) \qquad S^* &= U \cap \text{ support } (H^d \big|_S) \\ &= \{x \in U : H^d(S \cap B(x,r)) > 0 \text{ for all } r > 0\}. \end{aligned}$$

It is easy to check that S^* is always relatively closed in U, and that

$$(1.13) \qquad H^d(S \backslash S^*) = 0.$$

On the other hand, $S \backslash S^*$ could be nonempty, and in fact it could be any relatively closed subset of $U \backslash S^*$ with H^d-measure zero. This is because one can always add a relatively closed subset of H^d-measure zero to a quasiminimizing set without disturbing the quasiminimality property. Thus our results about the structure of quasiminimizing sets will only address the behavior of S^*.

The possibility of $S \backslash S^*$ being nonempty arises very naturally in certain variational problems. This is illustrated by the following scenario. Fix a (closed) solid torus T in $\mathbf{R}^3$. Given $\epsilon > 0$ (small), let Σ_ϵ denote the 2-dimensional surface $\{z \in \mathbf{R}^3 : \text{dist}(z, T) = \epsilon\}$. Set $U = \mathbf{R}^3 \backslash T$, and let g be a continuous positive function on $\mathbf{R}^3$. Consider the problem of finding minima for a functional of the form

$$(1.14) \qquad J_g(S) = \int_S g(x) dH^2(x),$$

where S ranges among compact subsets of U that satisfy $H^2(S) < \infty$ and which can be realized as the image $f(\Sigma_\epsilon)$ of a continuous mapping $f : \Sigma_\epsilon \to U$ which is homotopic to the standard embedding of Σ_ϵ (as mappings into U). (The precise choice of ϵ does not matter here, as long as it is sufficiently small.) For moderate choices of g one might expect to have minimizing S's that look roughly like a Σ_ϵ, but by making g be very large around T one can arrange to have minimizing S's which look like a sphere that surrounds T together with a single wire that goes through the hole in T. This wire is removed in the passage from S to S^*, and it is invisible for 2-dimensional Hausdorff measure, but it does reflect nontrivial topological information, about the way that S surrounds T.

In general, one can have minimizing sets (for suitable functionals) which behave like relatively-nice d-dimensional surfaces together with pieces of lower dimension which are more complicated than just a single wire. (See also Remark 11.26 and Chapter 12, especially Sections 12.4 and 12.5.)

REMARK 1.15. It is easy to formulate generalizations of Definition 1.9 to nonintegral dimensions d. However, the argument for Ahlfors-regularity given in Chapters 3–4 also shows that quasiminimizers for H^d have to have zero H^d-measure when d is not an integer.

CHAPTER 2

Uniform Rectifiability and the Main Result

Our main result is that if S is a quasiminimizer for H^d, then S^* is locally uniformly rectifiable, and contains big pieces of Lipschitz graphs locally. A more precise statement will be given soon, but first we explain the terminology which will be used.

DEFINITION 2.1. *A closed subset E of $\mathbf{R}^n$ is said to be* Ahlfors-regular *of dimension d if there is a constant C_0 such that*

$$(2.2) \qquad C_0^{-1} r^d \leq H^d\big(E \cap B(x,r)\big) \leq C_0 r^d$$

for all $x \in E$ and $0 < r < \operatorname{diam} E$.

DEFINITION 2.3. *Let $E \subseteq \mathbf{R}^n$ be an Ahlfors-regular set of dimension d. We say that E is* uniformly rectifiable *if there is a constant $C_1 > 0$ such that, for each choice of $x \in E$ and $0 < r < \operatorname{diam} E$, there exist a compact subset F of $E \cap B(x,r)$ and a mapping $f : F \to \mathbf{R}^d$ such that*

$$(2.4) \qquad H^d(F) \geq C_1^{-1} r^d$$

and

$$(2.5) \qquad C_1^{-1}|z - w| \leq |f(z) - f(w)| \leq C_1|z - w| \quad \text{for all } z, w \in F.$$

This is one of several equivalent characterizations of uniform rectifiability. In the language of [**8**], we are requiring that E have "big pieces of bilipschitz images" of subsets of $\mathbf{R}^d$ (BPBI). This should be compared with the ordinary notion of (countable) rectifiability, which is equivalent to asking that E be covered, except for a set of H^d-measure zero, by a countable union of sets which are bilipschitz equivalent to subsets of $\mathbf{R}^d$. (See [**12, 13, 20**]). The latter condition does not contain any information about E at definite scales, but only in limiting regimes of magnification at typical points. By contrast, uniform rectifiability provides quantitative information at all scales. It implies that E is contained in a subset of $\mathbf{R}^{n+1}$ with a fairly well-behaved (and global) parameterization by $\mathbf{R}^d$. It also implies that certain singular integral operators of Calderón-Zygmund type are bounded on $L^2(E, dH^d)$. See [**8**] for more information on uniform rectifiability, especially Chapter 1 of Part I.

In Definition 2.7 below we give a modest strengthening of uniform rectifiability.

DEFINITION 2.6. *By a C_2-Lipschitz graph of dimension d in $\mathbf{R}^n$ we mean the graph in $\mathbf{R}^n \simeq \mathbf{R}^d \times \mathbf{R}^{n-d}$ of a mapping $A : \mathbf{R}^d \to \mathbf{R}^{n-d}$ such that $|A(x) - A(y)| \leq C_2 |x - y|$ for all $x, y \in \mathbf{R}^d$, or the image of such a graph under a rotation of $\mathbf{R}^n$.*

DEFINITION 2.7. *Let $E \subseteq \mathbf{R}^n$ be an Ahlfors-regular set of dimension d. We say that E contains* big pieces of Lipschitz graphs *(or BPLG, for short) if there* is a constant $C_2 > 0$ such that, for each $x \in E$ and $0 < r < \operatorname{diam} E$, there is a C_2-Lipschitz graph Γ of dimension d in $\mathbf{R}^n$ which satisfies

(2.8) $$H^d\big(E \cap \Gamma \cap B(x,r)\big) \geq C_2^{-1} r^d.$$

This condition is known to be strictly stronger than uniform rectifiability (by an example of Hrycak), but the difference between the two is not enormous. We shall use later the fact that if E is (Ahlfors)-regular and uniformly rectifiable, then it contains big pieces of Lipschitz graphs if and only if it has "big projections", in the sense of the next definition. (See [7] for more information.)

DEFINITION 2.9. *An Ahlfors-regular set E of dimension d in $\mathbf{R}^n$ is said to have* big projections *if there is a constant $C_3 > 0$ such that for each $x \in E$ and $0 < r < \operatorname{diam} E$ there is a d-plane P with the property that*

(2.10) $$H^d\Big(\pi_P\big(E \cap B(x,r)\big)\Big) \geq C_3^{-1} r^d.$$

Here π_P denotes the orthogonal projection of $\mathbf{R}^n$ onto P.

With these definitions we can now state our main result about the structure of quasiminimizing sets.

THEOREM 2.11. *Let U be an open set in $\mathbf{R}^n$, and suppose that S is a (U, k, δ)-quasiminimizer for H^d. Let S^* be the support in U of the restriction of H^d to S, as in (1.12). Then for each $x \in S^*$ and radius R which satisfy*

(2.12) $$0 < R < \delta \quad \text{and} \quad B(x, 3R) \subseteq U,$$

there is a compact, Ahlfors-regular set E of dimension d such that

(2.13) $$S^* \cap B(x, R) \subseteq E \subseteq S^* \cap B(x, 2R)$$

and

(2.14)   *E is uniformly rectifiable and contains big pieces of Lipschitz graphs.*

The constants C_0, C_1, and C_2 for the regularity, uniform rectifiability, and BPLG condition for E can be taken to depend only on n and k (and not on S, x, R, or δ).

In the special case where $U = \mathbf{R}^n$ and $\delta = +\infty$, the conclusions of Theorem 2.11 simply say that S^* is Ahlfors-regular, uniformly rectifiable, and contains big pieces of Lipschitz graphs, with constants that depend only on k and n. In general we shall refer to the conclusions of Theorem 2.11 by saying that S^* is locally regular, locally uniformly rectifiable, and locally contains big pieces of Lipschitz graphs. Note that one might normally formulate the local versions of Ahlfors-regularity, uniform rectifiability, and the BPLG condition more directly, simply by localizing the original definitions. In the proof below we shall see that these more direct formulations imply the apparently stronger versions in Theorem 2.11.

The local Ahlfors-regularity of S^* was obtained by Almgren in [2], as well as ordinary (non-uniform) rectifiability properties. We shall re-prove it here, in part because we shall use the same construction to get Lipschitz mappings from S^* to $\mathbf{R}^d$ with controlled norm and substantial images (in terms of measure). This will be needed for the proof of the other parts of Theorem 2.11. Also, the upper bounds in the local Ahlfors-regularity condition in [2] have a constant C_0 which depends on the total mass $H^d(S)$ of S, and we prefer to avoid this. (Note that if one were taking multiplicities into account, then this would not work, e.g., S might be a d-plane with large (constant) multiplicities.)

Of course our conclusions about the behavior of S^* degenerate (or become more localized) as one approaches the complement of U, or as one goes to scales larger than δ. This corresponds to the restrictions on R in (2.12), and it is very natural given that the admissible deformations in the definition of a quasiminimizer are similarly restricted.

The proof of Theorem 2.11 will occupy us for the next several chapters.

REMARK 2.15. Let S, x, R, etc., be as in the statement of Theorem 2.11, and assume also that $R < \delta/2$. In this case one can say immediately that

$$\text{(2.16)} \qquad \operatorname{diam} S^* \cap B(x, R) \geq R,$$

and in fact that S^* intersects $\partial B(x, r)$ for each $r \in (0, R]$. To see this, suppose to the contrary that there is an $r \in (0, R]$ such that $S^* \cap \partial B(x, r) = \emptyset$. Because S^* is relatively closed in U, there must be an $\eta > 0$ such that

$$S^* \cap (\overline{B}(x, r) \setminus B(x, r - \eta)) = \emptyset.$$

(This also uses the fact that $\overline{B}(x, R) \subseteq U$, as in (2.12).) From here we get that

$$\text{(2.17)} \qquad H^d\bigl(S \cap (\overline{B}(x, r) \setminus B(x, r - \eta))\bigr) = 0,$$

by the definition of S^*. Once one has (2.17), it is easy to build a deformation $\phi : \mathbf{R}^n \to \mathbf{R}^n$ which fixes all points outside of $B(x, r)$, maps $S \cap B(x, r)$ to a set of H^d-measure 0, and satisfies the usual conditions (1.3)–(1.7). The quasiminimizing property of S then gives (1.8), which can be used to show that

$$H^d(S \cap B(x, r)) = 0.$$

This contradicts the assumption (from Theorem 2.11) that x lies in S^*. Thus we have that S^* intersects $\partial B(x, r)$, as claimed above.

CHAPTER 3

Lipschitz Projections into Skeleta

In some of the proofs below, it will be useful to have a systematic way of projecting d-dimensional sets (typically, pieces of quasiminimizers) into d-dimensional grids, or skeleta. A standard construction will be given in this chapter. We need to establish some notation first.

Given a closed cube Q in $\mathbf{R}^n$ and an integer $j \geq 0$, let $\Delta_j(Q)$ denote the collection of 2^{jn} (closed) subcubes of Q of diameter $2^{-j}\operatorname{diam}(Q)$ which are obtained from repeated dyadic subdivision of Q.

For each nonnegative integer $m \leq n$, let $\Delta_{j,m}(Q)$ denote the collection of all m-dimensional faces of cubes in $\Delta_j(Q)$. Thus each element of $\Delta_{j,m}(Q)$ is an m-dimensional (closed) cube in its own right, with sidelength equal to 2^{-j} times the sidelength of Q. Let $\mathcal{S}_{j,m}(Q)$ be the subset of Q which consists of all the points in the m-dimensional cubes in $\Delta_{j,m}(Q)$. We shall refer to $\mathcal{S}_{j,m}(Q)$ as the m-dimensional skeleton of order j in Q.

PROPOSITION 3.1. *Let $Q \subseteq \mathbf{R}^n$ be a closed cube, and let E be a compact subset of Q such that $H^d(E) < +\infty$. Then for each $j \geq 0$ there is a Lipschitz mapping $\phi : \mathbf{R}^n \to \mathbf{R}^n$ with the following properties:*

(3.2) $\qquad \phi(x) = x \quad \text{for} \quad x \in \mathbf{R}^n \backslash Q;$

(3.3) $\qquad \phi(x) = x \quad \text{for} \quad x \in \mathcal{S}_{j,d}(Q);$

(3.4) $\qquad \phi(E) \subseteq \mathcal{S}_{j,d}(Q) \cup \partial Q;$

(3.5) $\qquad \phi(R) \subseteq R \quad \text{for every} \quad R \in \Delta_j(Q);$

and

(3.6) $\qquad H^d\bigl(\phi(E \cap R)\bigr) \leq C H^d(E \cap R) \quad \text{for all} \quad R \in \Delta_j(Q),$

where C depends only on n and d (but not on j).

One should think of d as being strictly less than n here; if $d = n$, then we can simply take ϕ to be the identity mapping.

The type of construction used in the proof of Proposition 3.1 is quite common in geometric measure theory, as in "Federer-Fleming arguments." See Section 4.2 in [**13**], and the Deformation Theorem 4.2.9 on p406 of [**13**] in particular.

We see from (3.3) and (3.4) that ϕ behaves like a projection from E to the skeleton $\mathcal{S}_{j,d}(Q)$ inside Q. This is no longer true near ∂Q, because we want ϕ to equal the identity on the complement of Q. Notice also that we do not require any estimate on the Lipschitz norm of ϕ at this time.

Our mapping ϕ will be obtained as the last element of a finite sequence ϕ_n, $\phi_{n-1}, \ldots, \phi_d$ of Lipschitz mappings on $\mathbf{R}^n$. Each ϕ_m will satisfy the natural analogues of (3.2)–(3.6), with (3.3) and (3.4) replaced by

$$(3.7) \qquad \phi_m(x) = x \quad \text{for} \quad x \in \mathcal{S}_{j,m}(Q)$$

and

$$(3.8) \qquad \phi_m(E) \subseteq \mathcal{S}_{j,m}(Q) \cup \partial Q,$$

respectively.

We start with $\phi_n(x) = x$, which obviously satisfies the required conditions. Suppose now that $m > d$ is given, and that we already have $\phi_n, \ldots, \phi_m$ with the properties above. We want to define ϕ_{m-1} as

$$(3.9) \qquad \phi_{m-1} = \psi_{m-1} \circ \phi_m,$$

where ψ_{m-1} is a suitable Lipschitz mapping on $\mathbf{R}^n$.

LEMMA 3.10. *There is a Lipschitz mapping $\psi_{m-1} : Q \to Q$ such that*

$$(3.11) \quad \psi_{m-1}(x) = x \quad \text{for} \quad x \in \mathcal{S}_{j,m-1}(Q) \cup \partial Q,$$

$$(3.12) \quad \psi_{m-1}\big(\phi_m(E)\big) \subseteq \mathcal{S}_{j,m-1}(Q) \cup \partial Q,$$

$$(3.13) \quad \psi_{m-1}(R) \subseteq R \quad \text{for all} \quad R \in \Delta_j(Q),$$

and

$$(3.14) \qquad H^d\Big(\psi_{m-1}\big(\phi_m(E \cap R)\big)\Big) \leq C H^d\big(\phi_m(E \cap R)\big) \quad \text{for all} \quad R \in \Delta_j(Q).$$

Proposition 3.1 will follow immediately as soon as Lemma 3.10 is established. Indeed, once we have ψ_{m-1} as in Lemma 3.10, we can extend it to all of $\mathbf{R}^n$ by taking it to be the identity mapping outside of Q, and we can then define ϕ_{m-1} as in (3.9). This choice of ϕ_{m-1} will satisfy the required properties, because of the corresponding assertions for ϕ_m and ψ_{m-1}. In the end we take ϕ to be ϕ_d for Proposition 3.1, and this mapping has the desired features.

Let us now prove Lemma 3.10. We begin by choosing ψ_{m-1} on

$$\mathcal{S}_m = \mathcal{S}_{j,m}(Q) \cup \partial Q.$$

This will be the most important part of the definition of ψ_{m-1}, because (3.8) tells us that $\phi_m(E) \subseteq \mathcal{S}_m$. Notice that we must take $\psi_{m-1}(x) = x$ for all $x \in \mathcal{S}_{m-1}$, because of (3.11), and so our only choices concern the values of ψ_{m-1} on $\mathcal{S}_m \backslash \mathcal{S}_{m-1}$.

Denote by $\Delta^*_{j,m}(Q)$ the set of m-dimensional faces $T \in \Delta_{j,m}(Q)$ that are not contained in ∂Q. Thus $\mathcal{S}_m \backslash \mathcal{S}_{m-1}$ is the union of the "interiors" of faces $T \in \Delta^*_{j,m}(Q)$. We shall define ψ_{m-1} individually on each $T \in \Delta^*_{j,m}(Q)$, in such a way that

$$(3.15) \qquad \psi_{m-1}(x) = x \quad \text{on} \quad \partial T$$

(where ∂T denotes the $(m-1)$-dimensional boundary of T), and so that

$$(3.16) \qquad \psi_{m-1} : T \to T \quad \text{is Lipschitz}.$$

If we do this, then we shall have a consistent definition of ψ_{m-1} on all of $\mathcal{S}_m$ (because of (3.15) and (3.11)), and ψ_{m-1} will be Lipschitz on $\mathcal{S}_m$ because of the following.

OBSERVATION 3.17. *If f is a mapping on $\mathcal{S}_m$ such that the restriction of f to $\mathcal{S}_{m-1}$ is Lipschitz, and the restriction of f to each face $T \in \Delta_{j,m}^*(Q)$ is Lipschitz, then f is Lipschitz on $\mathcal{S}_m$, with a norm no greater than a bounded factor times the maximum of the Lipschitz norms of the restrictions of f to the T's in $\Delta_{j,m}^*(Q)$ and to $\mathcal{S}_{m-1}$.*

This follows from elementary geometric properties of $\mathcal{S}_m$, e.g., the fact that if x, y lie on two different faces of $\Delta_{j,m}^*(Q)$, then there is a path in $\mathcal{S}_m$ from x to y with length $\leq C|x-y|$ and which passes through $\mathcal{S}_{m-1}$.

Fix a $T \in \Delta_{j,m}^*(Q)$, and let us define ψ_{m-1} on T. We shall use the following type of "radial projection". Given any point $\xi \in \frac{1}{2}T$ (where $\frac{1}{2}T$ denotes the m-dimensional subcube of T with the same center but half the sidelength), define the projection $\theta_{\xi,T} : T \setminus \{\xi\} \to \partial T$ by the conditions

(3.18) $$\theta_{\xi,T}(x) \in \partial T \quad \text{and} \quad \theta_{\xi,T}(x) - \xi = \lambda(x-\xi)$$
$$\text{for some} \quad \lambda = \lambda(x, \xi, T) > 0$$

(for each $x \in T \setminus \{\xi\}$). Note that

(3.19) $$\theta_{\xi,T}(x) = x \quad \text{for} \quad x \in \partial T$$

and

(3.20) for each compact set $K \subseteq T \setminus \{\xi\}$, the restriction of $\theta_{\xi,T}$
to K is Lipschitz with norm $\leq C \operatorname{diam} T \operatorname{dist}(\xi, K)^{-1}$

This constant C depends on the dimension but not on K or T (or j, for that matter).

For each $T \in \Delta_{j,m}^*(Q)$, imagine that we select a point

$$\xi = \xi(T) \in \left(\frac{1}{2}T\right) \setminus \phi_m(E).$$

This is possible, because $H^d(\phi_m(E)) < \infty$ (since $H^d(E) < \infty$ by assumption, and ϕ_m is Lipschitz by induction hypothesis), and because $m > d$. We shall see later how to choose $\xi(T)$ efficiently for the size estimate (3.14), but for the moment let us just use the information that $\xi(T) \notin \phi_m(E)$ and see how we can get the rest of Lemma 3.10. Set

(3.21) $$\psi_{m-1}(x) = \theta_{\xi,T}(x) \quad \text{for} \quad x \in \bigl(T \cap \phi_m(E)\bigr) \cup \partial T.$$

This is compatible with (3.15) because of (3.19), and ψ_{m-1} is Lipschitz on

$$\bigl(T \cap \phi_m(E)\bigr) \cup \partial T$$

because of (3.20). (Remember from the statement of Proposition 3.1 that E is compact by assumption.) We can easily extend ψ_{m-1} as a Lipschitz mapping from T to T with a comparable norm, by standard extension results. (See [**13**], for instance.)

This completes our definition of ψ_{m-1} on each $T \in \Delta_{j,m}^*(Q)$. We already took ψ_{m-1} to be the identity mapping on $\mathcal{S}_{m-1} = \mathcal{S}_{j,m-1}(Q) \cup \partial Q$, and so we now have ψ_{m-1} defined on all of $\mathcal{S}_m$. Observation (3.17) implies that ψ_{m-1} is Lipschitz on $\mathcal{S}_m$, since (3.15) and (3.16) hold.

We still need to extend ψ_{m-1} to the rest of Q. We should be a little careful about this since we want $\psi_{m-1}(R)$ to be contained in R for each $R \in \Delta_{j,m}(Q)$. Notice that $\psi_{m-1}(R \cap \mathcal{S}_m) \subseteq R$ for every R, since $\psi_{m-1}(T) \subseteq T$ for all $T \in \Delta_{j,m}^*(Q)$

(and since ψ_{m-1} equals equals the identity mapping on ∂Q). Let us first describe how to extend ψ_{m-1} to $\mathcal{S}_{m+1}$. Because of Observation 3.17, it is enough to extend ψ_{m-1} to each face $T \in \Delta^*_{j,m+1}(Q)$ separately. This can be accomplished through the following construction. For each $T \in \Delta^*_{j,m+1}(Q)$, let c_T denote the center of T, and take $\psi_{m-1}(c_T)$ to be any point in the convex hull of $\psi_{m-1}(\partial T)$. We then define ψ_{m-1} on T through the requirement that it be linear on each line segment from c_T to ∂T. This gives a Lipschitz extension of ψ_{m-1} to T. By doing this for each $T \in \Delta^*_{j,m+1}(Q)$ we get a Lipschitz extension of ψ_{m-1} to $\mathcal{S}_{m+1}$. A useful feature of this method is that we have $\psi_{m-1}(R \cap \mathcal{S}_{m+1}) \subseteq R$ for all $R \in \Delta_j(Q)$, because of the corresponding property for $R \cap \mathcal{S}_m$. After repeating this construction several times, we obtain a Lipschitz extension of ψ_{m-1} to Q which satisfies (3.13). In particular, ψ_{m-1} maps Q to Q.

Note that ψ_{m-1} satisfies (3.11), and in fact this requirement was incorporated into the initial steps of the definition of ψ_{m-1} (shortly after the statement of Lemma 3.10). To get (3.12), we first use (3.8) to know that every element x of $\phi_m(E)$ lies in at least one of $\mathcal{S}_{j,m}(Q)$ and ∂Q. If x lies in ∂Q, then $\psi_{m-1}(x) = x$, by (3.11), and this is fine for (3.12). If $x \in \mathcal{S}_{j,m}(Q) \setminus \partial Q$, then $x \in T$ for some $T \in \Delta^*_j(Q)$. In this case we apply (3.21) and (3.18) to conclude that $\psi_{m-1}(x) \in \partial T$, which is also adequate for (3.12). In order to complete the proof of Lemma 3.10 (and hence of Proposition 3.1), it suffices to show that we can choose the points $\xi(T)$, $T \in \Delta^*_{j,m}(Q)$, so that (3.14) holds. To do this we shall use the following.

LEMMA 3.22. *Let T be an m-dimensional cube in $\mathbf{R}^m$, and let $F \subseteq T$ be a closed set such that $H^d(F) < +\infty$ for some $d < m$. Then*

$$(3.23) \qquad \int_{\xi \in (\frac{1}{2}T) \setminus F} H^d\big(\theta_{\xi,T}(F)\big) d\xi \leq C (\operatorname{diam} T)^m H^d(F),$$

where C depends on m and d, but nothing else.

Note that the "$d\xi$" on the left side of (3.23) refers to m-dimensional Lebesgue measure on T. Also, one need not worry about the measurability of the integrand on the left side of (3.23), as the proof will show that the inequality holds with the left side interpreted as an "outer" integral, which is fine for our purposes.

For a given point ξ, a simple computation (using (3.20)) shows that

$$(3.24) \qquad H^d\big(\theta_{\xi,T}(F)\big) \leq C \int_F |x - \xi|^{-d} (\operatorname{diam} T)^d dH^d(x).$$

We can integrate this over $(\frac{1}{2}T) \setminus F$ and apply the Fubini theorem to get that

$$(3.25) \qquad \int_{(\frac{1}{2}T) \setminus F} H^d\big(\theta_{\xi,T}(F)\big) d\xi \leq C (\operatorname{diam} T)^d \int_F \int_T |x - \xi|^{-d} d\xi dH^d(x).$$

The ξ-integral converges because $d < m$, and its value is $\leq C(\operatorname{diam} T)^{m-d}$. This implies (3.23), as desired.

Let us return to the proof of (3.14). Let $T \in \Delta^*_{j,m}(Q)$ be given, and denote by $\Delta(T)$ the set of cubes $R \in \Delta_j(Q)$ which contain T (as an m-dimensional face). For each $R \in \Delta(T)$, put $F_{R,T} = \phi_m(E \cap R) \cap T$. These are sets of m-dimensional Lebesgue measure zero, since ϕ_m is Lipschitz, $H^d(E) < +\infty$, and $m > d$. Thus

$\left(\frac{1}{2}T\right)\setminus\bigcup_{R\in\Delta(T)} F_{R,T}$ has the same measure as $\frac{1}{2}T$. By Lemma 3.22 and the Tchebychev inequality, there is a point $\xi = \xi(T)$ in $\left(\frac{1}{2}T\right)\setminus\bigcup_{\Delta(T)} F_{R,T}$ such that

$$(3.26) \qquad H^d\big(\theta_{\xi,T}(F_{R,T})\big) \leq CH^d(F_{R,T}) \quad \text{for all } R \in \Delta(T).$$

This also uses the fact that there are only boundedly many R's in $\Delta(T)$. We choose $\xi = \xi(T)$ in this way for every T in $\Delta_{j,m}^*(Q)$, and we want to check now that (3.14) holds (with these choices).

Let $R \in \Delta_j(Q)$ be given, and set $F = \phi_m(E \cap R)$. We want to show that $H^d\big(\psi_{m-1}(F)\big) \leq CH^d(F)$ for a suitable constant C. From (3.8) we have that $F \subseteq \mathcal{S}_{j,m}(Q) \cup \partial Q$, so that

$$(3.27) \qquad F \subseteq \big(F \cap (\mathcal{S}_{j,m-1}(Q) \cup \partial Q)\big) \cup \Big(\bigcup_{T \in \Delta_{j,m}^*(Q)} F \cap \text{int}(T)\Big),$$

where int(T) denotes the interior of T. For the first part we have that

$$H^d\big(\psi_{m-1}\big(F \cap (\mathcal{S}_{j,m-1}(Q) \cup \partial Q)\big)\big) = H^d\big(F \cap (\mathcal{S}_{j,m-1}(Q) \cup \partial Q)\big) \leq H^d(F),$$

since ψ_{m-1} equals the identity mapping on $\mathcal{S}_{j,m-1}(Q) \cup \partial Q$, as in (3.11). To treat the remaining portion, we use the analogue of (3.5) for ϕ_m to say that $F \subseteq R$, so that F intersects int(T) for some T in $\Delta_{j,m}^*(Q)$ only when T is contained in R. In other words, we should have $R \in \Delta(T)$ in this case. For each of these T's we have that

$$(3.28) \qquad H^d\big(\psi_{m-1}(F \cap T)\big) = H^d\big(\psi_{m-1}(F_{R,T})\big) \leq CH^d(F_{R,T}) \leq CH^d(F)$$

by (3.26) (and (3.21)). To get the desired estimate for $H^d\big(\psi_{m-1}(F)\big)$, we sum (3.28) over the T's in $\Delta_{j,m}^*(Q)$ which are contained in R (of which there are only boundedly many), and we combine the result with the earlier (trivial) estimate for $H^d\big(\psi_{m-1}\big(F \cap (\mathcal{S}_{j,m-1}(Q) \cup \partial Q)\big)\big)$.

This completes the proof of (3.14), and hence of Lemma 3.10. As observed before, Proposition 3.1 now follows as well.

We shall need later a variant of Proposition 3.1, in which we assume that E is "semi-regular of dimension d" (defined below), and we get a bound for the Lipschitz norm of the mapping ϕ in Proposition 3.1.

DEFINITION 3.29. *A closed subset E of $\mathbf{R}^n$ is said to be* semi-regular of dimension d *(with constant C) if for each $x \in \mathbf{R}^n$ and each choice of radius $0 < r \leq R$ the set $E \cap B(x,R)$ can be covered by a collection of at most $Cr^{-d}R^d$ balls of radius r.*

Ahlfors-regular sets are always semi-regular of the same dimension. To see this, assume that E is regular, and let x, R, and r be as in Definition 3.29. Let $\{y_i\}_{i \in I}$ be a maximal family of points in $E \cap B(x,R)$ which lie at distance $\geq r$ from each other. The maximality of this family ensures that $E \cap B(x,R)$ is covered by the balls $B(y_i, r)$, and so it is enough to show that I has no more than $Cr^{-d}R^d$

elements. Indeed, we have that

$$\#I = r^{-d} \sum_{i \in I} r^d \leq Cr^{-d} \sum_{i \in I} H^d\big(E \cap B(y_i, \tfrac{r}{2})\big) \tag{3.30}$$

$$\leq Cr^{-d} H^d\Big(E \cap \big(\bigcup_{i \in I} B(y_i, \tfrac{r}{2})\big)\Big)$$

$$\leq Cr^{-d} H^d\big(E \cap B(x, 2R)\big) \leq Cr^{-d} R^d.$$

The second inequality uses the fact that the balls $B(y_i, \tfrac{r}{2})$ $i \in I$, are pairwise disjoint, since $|y_i - y_j| \geq r$ when $i \neq j$ by construction.

LEMMA 3.31. *Let Q, E and $j \geq 0$ be as in Proposition 3.1, and suppose in addition that E is semi-regular of dimension d. Then we can find a Lipschitz mapping $\phi : \mathbf{R}^n \to \mathbf{R}^n$ with the same properties as in Proposition 3.1 (i.e., (3.2)–(3.6)), and which in addition has Lipschitz norm bounded by a constant that depends only on n, d, and the semi-regularity constant for E.*

To prove this lemma, we employ exactly the same construction as in Proposition 3.1, except that we modify slightly the selection of the points $\xi(T)$ for $T \in \Delta_{j,m}^*(Q)$. As before, we argue by induction, and thus we assume that we already have functions $\phi_n, \ldots, \phi_m$ which satisfy (3.2)–(3.5), $m > d$, and that ϕ_m is C_m-Lipschitz for some constant C_m which depends only on n, d, m, and the semi-regularity constant for E. We want to show that we can choose the points $\xi(T)$, $T \in \Delta_{j,m}^*(T)$, in the construction above so that ψ_{m-1} is C-Lipschitz, where C is permitted to depend on n, d, m, the semi-regularity constant for E, and C_m. Once we do this, Lemma 3.31 will follow, for the same reasons as before.

SUBLEMMA 3.32. *For each $T \in \Delta_{j,m}^*(Q)$ and each $r < 2^{-j} \operatorname{diam} Q$,*

$$\phi_m(E) \cap T$$

can be covered by $\leq Cr^{-d}[2^{-j} \operatorname{diam} Q]^d$ balls of radius r, where C depends only on n, C_m, and the semi-regularity constant for E.

To see this, let R denote the union of the cubes in $\Delta_j(Q)$ which intersect T. Then $\phi_m(E) \cap T \subseteq \phi_m(E \cap R)$, because ϕ_m satisfies (3.5). Our assumption of semi-regularity for E implies that we can cover $E \cap R$ with $\leq CC_m^d r^{-d}[2^{-j} \operatorname{diam} Q]^d$ balls of radius $C_m^{-1} r$. Therefore, we can cover $\phi_m(E \cap R)$ by the same number of balls of radius r, because ϕ_m is C_m-Lipschitz. This proves Sublemma 3.22.

Fix $T \in \Delta_{j,m}^*(Q)$, and let us apply Sublemma 3.32, with

$$r = \alpha_m 2^{-j} \operatorname{diam} Q,$$

where $\alpha_m > 0$ is a small constant which will be chosen soon. Thus we obtain that $\phi_m(E) \cap T$ can be covered by a collection $\{B_i\}$ of $\leq C\alpha_m^{-d}$ balls of radius r. On the other hand, $\tfrac{1}{2}T$ cannot be covered by fewer than $A^{-1}\alpha_m^{-m}$ balls of radius $2r$, at least if the constant A is chosen large enough (depending on m but not on T, Q, or j). This is because T is an m-dimensional cube with sidelength equal to 2^{-j} times the sidelength of Q. Since $m > d$, we can choose α_m small enough so that

$$C\alpha_m^{-d} < A^{-1}\alpha_m^{-m},$$

and where α_m depends only on n, m, and the semi-regularity constant for E. With this choice of α_m we have that the balls $2B_i$ cannot cover $\tfrac{1}{2}T$, and hence that there is a point $\xi = \xi(T) \in \frac{1}{2}T$ such that

(3.33) $$\operatorname{dist}(\xi, \phi_m(E) \cap T) \geq \alpha_m 2^{-j} \operatorname{diam} Q.$$

We do this for each $T \in \Delta_{j,m}^*(Q)$, and then construct ψ_{m-1} as in the proof of Proposition 3.1 (with these choices of $\xi(T)$). An important point now is that the restriction of $\theta_{\xi,T}$ to $(\phi_m(E) \cap T) \cup \partial T$ is $C\alpha_m^{-1}$-Lipschitz, where C depends on the dimension n but not on T, Q, E, etc. This follows from (3.20) and (3.33). Once we have this, the same series of extensions can be used as before to get a mapping $\psi_{m-1} : \mathbf{R}^n \to \mathbf{R}^n$ which now has bounded Lipschitz norm. This permits us to define ϕ_{m-1} as in (3.9), with a bound on its Lipschitz norm as well. The rest of the argument is exactly the same as before, i.e., one stops if $m - 1 = d$, and otherwise one repeats the procedure in order to define ϕ_{m-2}. This completes the proof of Lemma 3.31.

CHAPTER 4

Local Ahlfors-Regularity

In this chapter we establish the local Ahlfors-regularity of S^* when S is a quasiminimizer for H^d.

PROPOSITION 4.1. *Let U be an open set in $\mathbf{R}^n$, and let S be a (U, k, δ)-quasiminimizer for H^d. If Q is a cube centered on S^* which satisfies*

(4.2) $$2Q \subseteq U \quad \text{and} \quad \operatorname{diam} Q < \delta/2,$$

then

(4.3) $$C^{-1}(\operatorname{diam} Q)^d \leq H^d(S \cap Q) \leq C(\operatorname{diam} Q)^d,$$

where C depends only on n and k.

This is only a small improvement of Theorem II.3 on pp. 54–55 of [**2**], the main difference being that upper bound given in [**2**] depends also on the total mass $H^d(S)$.

The proof of the two inequalities in (4.3) will be similar to each other. In both cases we shall employ Proposition 3.1 to produce a deformation ϕ, which will be used to test the quasiminimality of S. This will lead to inequalities which relate the mass of S in Q and the mass of S in a thin shell near the boundary of Q. These inequalities will then be used like differential inequalities to get the desired estimates.

Let S be a (U, k, δ)-quasiminimizer, and let Q be a closed cube such that $Q \subseteq U$ and $\operatorname{diam} Q < \delta$. (This is not necessarily the same cube as in the statement of Proposition 4.1.) Also let $j \geq 0$ be given, and apply Proposition 3.1 to the set $E = S \cap Q$. (Note that $H^d(E) < +\infty$, because of (1.2).) Thus we get a Lipschitz mapping $\phi : \mathbf{R}^n \to \mathbf{R}^n$ with certain properties, and we want to apply the definition of quasiminimality with this choice of ϕ.

Set $W = \{x \in \mathbf{R}^n : \phi(x) \neq x\}$. Then

(4.4) $$W \cup \phi(W) \subseteq Q,$$

by (3.2) and (3.5), so that ϕ satisfies the requirements (1.5) and (1.6) trivially. It also satisfies (1.7) as in Remark 1.10. From (1.8) we obtain that

(4.5) $$H^d(S \cap W) \leq k H^d\big(\phi(S \cap W)\big).$$

Let $A_j(Q)$ denote the union of the cubes in $\Delta_j(Q)$ which intersect ∂Q. (Here $\Delta_j(Q)$ is as in Chapter 3.) Thus

(4.6) $$A_j(Q) = Q \backslash \operatorname{int}\big((1 - 2^{-j+1})Q\big)$$

if $j \geq 1$ (where "int Y" denotes the interior of the set Y). Also set

(4.7) $$H_j(Q) = Q \backslash A_j(Q) = \operatorname{int}\big((1 - 2^{-j+1})Q\big).$$

4. LOCAL AHLFORS-REGULARITY

To prove the second inequality in (4.3), we want to get an upper bound for $H^d(S \cap H_j(Q))$. Notice first that

$$\begin{aligned}
(4.8) \qquad H^d(S \cap \mathcal{S}_{j,d}(Q)) &\leq H^d(\mathcal{S}_{j,d}(Q)) \\
&\leq C 2^{(n-d)j} (\operatorname{diam} Q)^d.
\end{aligned}$$

The rest of $S \cap H_j(Q)$ lies in W because of (3.4). Thus (4.5) and (4.8) imply that

$$(4.9) \qquad H^d(S \cap H_j(Q)) \leq k H^d(\phi(S \cap W)) + C 2^{(n-d)j}(\operatorname{diam} Q)^d.$$

Since $S \cap W \subseteq S \cap Q = E$, we can use (3.4), (3.5), and (3.6) to obtain that

$$\begin{aligned}
(4.10) \qquad H^d(\phi(S \cap W)) &\leq H^d(\mathcal{S}_{j,d}(Q)) + H^d(\partial Q \cap \phi(S \cap W)) \\
&\leq C 2^{(n-d)j}(\operatorname{diam} Q)^d + H^d(\phi(S \cap A_j(Q))) \\
&\leq C 2^{(n-d)j}(\operatorname{diam} Q)^d + C H^d(S \cap A_j(Q)).
\end{aligned}$$

Therefore

$$(4.11) \qquad H^d(S \cap H_j(Q)) \leq C k H^d(S \cap A_j(Q)) + C 2^{(n-d)j}(\operatorname{diam} Q)^d.$$

Note that (4.11) holds for all choices of cubes $Q \subseteq U$ and integers $j \geq 1$, under merely the condition that $\operatorname{diam} Q < \delta$. This is all that we shall need in order to prove the second inequality in (4.3).

Now let Q_0 be any cube such that $2Q_0 \subseteq U$ and $\operatorname{diam} Q_0 < \delta/2$, as in (4.2). We want to prove that

$$m_0 = (\operatorname{diam} Q_0)^{-d} H^d(S \cap Q_0)$$

is not too large. Let us proceed by contradiction, assuming that m_0 is enormous, and trying to contradict the assumption (1.2). We shall do this by constructing a slowly-increasing sequence of cubes $Q_i \subseteq 2Q_0$ such that the masses $H^d(S \cap Q_i)$ increase geometrically (and so that $H^d(S \cap 2Q_0) = +\infty$ in the end).

We start by taking $Q_i = Q_0$ when $i = 0$. Assume now that $Q_0, \ldots, Q_i$ have already been constructed, with $Q_0 \subseteq Q_\ell \subseteq 2Q_0$ for $\ell = 0, 1, \ldots, i$, and let us choose Q_{i+1}. Set

$$(4.12) \qquad m_i = (\operatorname{diam} Q_0)^{-d} H^d(S \cap Q_i),$$

and let $j = j(i)$ be the integer which satisfies

$$(4.13) \qquad m_i 2^{-(n-d)} \leq 2^{d+1} C 2^{(n-d)j} < m_i,$$

where C is the same constant as in (4.11). Note that $m_i \geq m_0$, since $Q_0 \subseteq Q_i$; if m_0 is large enough (depending on d and n), then $j(i)$ must be at least 10. We shall use this freely in the arguments that follow. Put

$$(4.14) \qquad Q_{i+1} = (1 + 2^{-j+2}) Q_i.$$

From the definition (4.7) of $H_j(Q)$ we have that

$$(4.15) \qquad Q_i \subseteq H_{j(i)}(Q_{i+1}),$$

as one can easily check. Let us assume for the moment that $Q_{i+1} \subseteq 2Q_0$ and derive some estimates. (Note that we automatically have $Q_{i+1} \subseteq 2Q_0$ for the first few values of i, at least if m_0 is large enough, because $j(i)$ is then pretty large.)

Since we are assuming that $Q_{i+1} \subseteq 2Q_0$, we can apply (4.11) with $Q = Q_{i+1}$ and $j = j(i)$. Using this, (4.12), and (4.15) we get that

$$\begin{aligned}(4.16) \quad m_i(\operatorname{diam} Q_0)^d &\leq H^d\big(H_{j(i)}(Q_{i+1}) \cap S\big) \\ &\leq CkH^d\big(S \cap A_{j(i)}(Q_{i+1})\big) + C2^{(n-d)j(i)}(2\operatorname{diam} Q_0)^d.\end{aligned}$$

The second inequality in (4.13) implies that the last term in (4.16) is less than $\frac{1}{2}m_i(\operatorname{diam} Q_0)^d$. Therefore, (4.16) can be replaced with

$$(4.17) \qquad m_i(\operatorname{diam} Q_0)^d \leq 2CkH^d\big(S \cap A_{j(i)}(Q_{i+1})\big).$$

This permits us to conclude that

$$\begin{aligned}(4.18) \\ m_{i+1} &= (\operatorname{diam} Q_0)^{-d} H^d(S \cap Q_{i+1}) \\ &= (\operatorname{diam} Q_0)^{-d} H^d\big(H_{j(i)}(Q_{i+1}) \cap S\big) + (\operatorname{diam} Q_0)^{-d} H^d\big(S \cap A_{j(i)}(Q_{i+1})\big) \\ &\geq m_i + (\operatorname{diam} Q_0)^{-d} H^d\big(S \cap A_{j(i)}(Q_{i+1})\big) \\ &\geq m_i\left(1 + \frac{1}{2Ck}\right)\end{aligned}$$

(where the first inequality is the same as the first inequality in (4.16), and the second inequality uses (4.17)). Notice that the expansion factor of $1 + 1/2Ck$ on the right-hand side does not depend on i.

Set $\rho = 1 + \frac{1}{2Ck} > 1$. Of course the Q_ℓ's for $\ell \leq i$ should be chosen in exactly the same manner as Q_{i+1}, so that

$$(4.19) \qquad m_\ell \geq \rho^\ell m_0 \quad \text{for } 0 \leq \ell \leq i+1,$$

by (4.18) and its counterpart for the Q_ℓ's, $\ell < i$. Using (4.14) and (4.13) we get that

$$\begin{aligned}(4.20) \qquad \log\left(\frac{\operatorname{diam} Q_{i+1}}{\operatorname{diam} Q_0}\right) &= \sum_{\ell=0}^{i} \log(1 + 2^{-j(\ell)+2}) \\ &\leq \sum_{\ell=0}^{i} 2^{-j(\ell)+2} \\ &\leq C \sum_{\ell=0}^{i} (m_\ell)^{\frac{-1}{n-d}} \\ &\leq C m_0^{-\frac{1}{n-d}} \sum_{\ell=0}^{\infty} \rho^{\frac{-\ell}{n-d}} \\ &\leq C' m_0^{-\frac{1}{n-d}}.\end{aligned}$$

If m_0 is large enough (depending only on n and k), then this implies that

$$\operatorname{diam} Q_{i+1} < \frac{3}{2} \operatorname{diam} Q_0.$$

These computations were made under the assumption that $Q_{i+1} \subseteq 2Q_0$, but we can use the same method to prove that this assumption holds. More precisely, the same analysis as in (4.20) implies that $\operatorname{diam} Q_i < \frac{3}{2} \operatorname{diam} Q_0$, and therefore

$$\operatorname{diam} Q_{i+1} = \left(1 + 2^{-j(i)+2}\right) \operatorname{diam} Q_i$$
$$< \left(1 + 2^{-j(i)+2}\right) \frac{3}{2} \operatorname{diam} Q_0$$

by (4.14). Since $j(i) \geq 10$ we get that $\operatorname{diam} Q_{i+1} < 2 \operatorname{diam} Q_0$, so that $Q_{i+1} \subseteq 2Q_0$, as desired.

To summarize, we have shown that if $Q_0, \ldots, Q_i$ are chosen in the manner described above, with $Q_0 \subseteq Q_\ell \subseteq 2Q_0$ for $0 \leq \ell \leq i$, then we can define Q_{i+1} as in (4.14), and we also have that $Q_0 \subseteq Q_{i+1} \subseteq 2Q_0$. Thus we may repeat the process to get an infinite sequence $\{Q_i\}_{i=0}^{\infty}$ of cubes with the same properties as above.

In particular, for each $i \geq 0$ we have that

$$(4.21) \qquad H^d(S \cap 2Q_0) \geq H^d(S \cap Q_i) \geq (\operatorname{diam} Q_0)^d \rho^i m_0,$$

because of (4.19). Thus $H^d(S \cap 2Q) = +\infty$, which contradicts (1.2). This completes the proof of the second inequality in (4.3) for any cube Q_0 that satisfies (4.2). (Note that we do not need to know that Q_0 is centered on S^* for this part of (4.3).)

Next we prove the first inequality in (4.3). Let Q be a cube contained in U which satisfies $\operatorname{diam} Q < \delta$, and let j be any positive integer (for which we shall make a particular choice later). We apply Proposition 3.1 with $E = S \cap Q$ to get a Lipschitz mapping ϕ, but we shall not use the quasiminimality of S directly with this choice of ϕ. Let us assume for the time being that

$$(4.22) \qquad H^d(S \cap Q) \leq C^{-1} \beta 2^{-jd} (\operatorname{diam} Q)^d,$$

where C is the constant in (3.6), and β is a positive number which depends only on n and d and which is about to be specified. (If (4.22) were not true, then we would get a lower bound for $H^d(S \cap Q)$, albeit one that depends on j.)

Let T be any one of the d-dimensional cubes in $\Delta_{j,d}(Q)$ of which $\mathcal{S}_{j,d}(Q)$ is composed. (Recall that $\Delta_{j,d}(Q)$ and $\mathcal{S}_{j,d}(Q)$ were defined as in Chapter 3.) We want to choose β small enough so that $\phi(S \cap Q)$ cannot wholly contain $\frac{1}{2}T$ for any such T. If $s(Q)$ denotes the sidelength of Q, then

$$H^d\left(\tfrac{1}{2}T\right) = 2^{-(j+1)d} s(Q)^d,$$

except perhaps for a normalizing factor (that depends only on d). If β is small enough, depending only on n and d, then

$$H^d\left(\phi(S \cap Q) \cap T\right) < H^d\left(\tfrac{1}{2}T\right)$$

for each $T \in \Delta_{j,d}(Q)$, because of (4.22) and (3.6). (More precisely, one should apply (3.6) to each of the cubes R in $\Delta_j(Q)$ which contain T as a face.) We fix β once and for all so that this is true, and we therefore have that $\phi(S \cap Q)$ cannot contain $\frac{1}{2}T$ for any T in $\Delta_{j,d}(Q)$.

For each T in $\Delta_{j,d}(Q)$ that is not contained in ∂Q, choose

$$\xi \in \tfrac{1}{2}T \backslash \phi(S \cap Q),$$

and let $\theta = \theta_{\xi,T} : T \cap \phi(S \cap Q) \to \partial T$ be the same kind of radial projection as in (3.18). It is easy to extend θ to a Lipschitz mapping $\theta : T \to T$ such $\theta(x) = x$ on ∂T. When $T \in \Delta_{j,d}(Q)$ is contained in ∂Q, simply set $\theta(x) = x$ on T. Note that all the mappings θ that we just defined are equal to the identity on the $(d-1)$-skeleton $\mathcal{S}_{j,d-1}(Q)$. By Observation 3.17, we can combine all these local choices to get a single Lipschitz mapping on $\mathcal{S}_{j,d}(Q)$, which we shall also call θ. We can extend θ further to a Lipschitz mapping from $\mathbf{R}^n$ to itself that satisfies $\theta(x) = x$ when $x \in \mathbf{R}^n \backslash Q$ and $\theta(Q) \subseteq Q$. This is analogous to some steps in the proof of Lemma 3.10 (a couple of paragraphs after (3.21)).

Set $\widetilde{\phi} = \theta \circ \phi$ and $W = \{x \in \mathbf{R}^n : \widetilde{\phi}(x) \neq x\}$. It is clear that $\widetilde{\phi} : \mathbf{R}^n \to \mathbf{R}^n$ is Lipschitz and that $W \cup \widetilde{\phi}(W) \subseteq Q$ (because of (3.2) and (3.5), and the facts that $\theta(x) = x$ on $\mathbf{R}^n \backslash Q$ and $\theta(Q) \subseteq Q$). Thus $\widetilde{\phi}$ satisfies the requirements (1.3)–(1.7) (using Remark 1.10 for the latter), and (1.8) yields

$$(4.23) \qquad H^d(S \cap W) \leq k H^d(\widetilde{\phi}(S \cap W)).$$

Note that $\theta(\phi(S \cap Q) \backslash \partial Q)$ is contained in the $(d-1)$-dimensional skeleton $\mathcal{S}_{j,d-1}(Q)$. In particular, if $H_j(Q)$ is as in (4.7), then

$$\widetilde{\phi}(S \cap H_j(Q)) \subseteq \mathcal{S}_{j,d-1}(Q).$$

This implies that H^d-almost every point in $S \cap H_j(Q)$ lies in W, since $H^d(\mathcal{S}_{j,d-1}(Q))$ is 0. Also,

$$\widetilde{\phi}(S \cap Q) = \theta(\phi(S \cap Q))$$
$$\subseteq [\partial Q \cap \phi(S \cap Q)] \cup \mathcal{S}_{j,d-1}(Q),$$

because of the properties of ϕ and the way that we chose θ. More precisely, $\phi(S \cap Q)$ is contained in $\partial Q \cup \mathcal{S}_{j,d}(Q)$ as in (3.4), and then θ fixes every point in ∂Q while pushing points in $\phi(S \cap Q)$ which are not in ∂Q into $\mathcal{S}_{j,d-1}(Q)$, by construction. Using these observations and (4.23) we obtain that

$$(4.24) \qquad H^d(S \cap H_j(Q)) \leq H^d(S \cap W)$$
$$\leq k H^d(\widetilde{\phi}(S \cap W)) \leq k H^d(\widetilde{\phi}(S \cap Q))$$
$$\leq k H^d(\partial Q \cap \phi(S \cap Q)).$$

On the other hand, (3.5) implies that if $x \in S \cap Q$ is mapped into ∂Q by ϕ, then $x \in A_j(Q)$, where $A_j(Q)$ is as in (4.6). From this and (3.6) we conclude that

$$H^d(\partial Q \cap \phi(S \cap Q)) \leq C H^d(S \cap A_j(Q)).$$

More precisely, we apply (3.6) to the cubes $R \in \Delta_j$ of which $A_j(Q)$ is composed, and then sum over R. Combining this with (4.24) yields

$$(4.25) \qquad H^d(S \cap H_j(Q)) \leq C k H^d(S \cap A_j(Q)).$$

Let us summarize a bit. If Q is a cube in U with $\operatorname{diam} Q < \delta$, and if j is a positive integer such that (4.22) holds, then we have shown that (4.25) must be true. Keep in mind that C and β in (4.22) depend only on n and d.

We are now ready to prove the lower bound in (4.3). Let Q_0 be a cube contained in U such that $\operatorname{diam} Q_0 < \delta$. We want to prove that if

$$m_0 = (\operatorname{diam} Q_0)^{-d} H^d(S \cap Q_0)$$

is small enough, then $H^d(S \cap (\frac{1}{2} Q_0)) = 0$. This would imply that Q_0 was not centered on S^*, as required in Proposition 4.1. To this end, we shall define a slowly decreasing sequence of cubes Q_i, $i \geq 1$, which satisfy

$$(4.26) \qquad \frac{1}{2} Q_0 \subseteq Q_i \subseteq Q_0$$

for all i, and such that
$$m_i = (\operatorname{diam} Q_0)^{-d} H^d(S \cap Q_i)$$
decreases geometrically.

Suppose that we have already selected Q_i so that (4.26) holds, and let us choose Q_{i+1}. First we define $j(i)$ to be the integer that satisfies
$$(4.27) \qquad m_i \leq C^{-1} \beta 2^{-(j(i)+1)d} < 2^d m_i,$$
where C and β are as in (4.22). If m_0 is sufficiently small, then $m_i \leq m_0$ is small too, and $j(i)$ must be ≥ 10. We shall use this freely in the arguments that follow. Set
$$(4.28) \qquad Q_{i+1} = \left(1 - 2^{-j(i)+2}\right) Q_i.$$
Clearly $Q_{i+1} \subseteq H_{j(i)}(Q_i)$. (See (4.7) for the definition of $H_j(Q)$.) Also, Q_i satisfies (4.22) with $j = j(i)$, because of (4.26) and the first half of (4.27). Thus we can apply (4.25) with $Q = Q_i$ to obtain
$$(4.29) \qquad \begin{aligned} H^d(S \cap Q_{i+1}) &\leq H^d(S \cap H_{j(i)}(Q_i)) \\ &\leq Ck H^d(S \cap A_{j(i)}(Q_i)) \\ &\leq Ck H^d(S \cap (Q_i \setminus Q_{i+1})). \end{aligned}$$
(The last inequality uses (4.6).) In other words, $m_{i+1} \leq Ck\,(m_i - m_{i+1})$, and so $m_{i+1} \leq \left(1 + \frac{1}{Ck}\right)^{-1} m_i$.

Of course we have a similar inequality for $m_{\ell+1}$ and m_ℓ when $\ell < i$ (assuming that the Q_ℓ's for $\ell \leq i$ were chosen in the same manner as for Q_{i+1}). This implies that
$$(4.30) \qquad m_\ell \leq \rho^\ell m_0 \quad \text{for } \ell \leq i + 1,$$
where $\rho = \left(1 + \frac{1}{Ck}\right)^{-1}$ (which satisfies $\rho < 1$ this time).

The idea that the Q_j's decrease slowly is captured by the following inequalities:
$$(4.31) \qquad \begin{aligned} \log\left(\frac{\operatorname{diam} Q_0}{\operatorname{diam} Q_{i+1}}\right) &= \sum_{\ell=0}^{i} -\log(1 - 2^{-j(\ell)+2}) \\ &\leq \sum_{\ell=0}^{i} 2^{-j(\ell)+3} \\ &\leq C' \sum_{\ell=0}^{i} m_i^{1/d} \leq C' m_0^{1/d} \sum_{\ell=0}^{\infty} \rho^{\ell/d} \\ &\leq C'' m_0^{1/d}. \end{aligned}$$
Here we have used (4.28), (4.27), (4.30), and the fact that $j(\ell) \geq 10$ for all $\ell \leq i$. Note that the constants C', C'' here depend only on n and k. If m_0 is small enough (depending only on n and k), then (4.31) implies that $Q_{i+1} \supseteq \frac{1}{2} Q_0$. Thus Q_{i+1} also satisfies (4.26), and we can repeat the process. In the end we get a sequence of cubes Q_i which satisfy both (4.26) and (4.30), and therefore
$$(4.32) \qquad H^d\left(S \cap \left(\tfrac{1}{2} Q_0\right)\right) \leq H^d(S \cap Q_i) \leq \rho^i H^d(S \cap Q_0)$$

for all $i \geq 0$. (This uses also the definition of m_i as $(\operatorname{diam} Q_0)^{-d} H^d(S \cap Q_i)$.) Thus we conclude that $H^d\bigl(S \cap (\frac{1}{2}Q_0)\bigr) = 0$. In particular, Q_0 cannot be centered on S^*, the support of H^d restricted to S, if m_0 is sufficiently small (in a way that depends only on n and k). This completes the proof of the lower bound in (4.3). We established the upper bound in (4.3) already, and so the proof of Proposition 4.1 is now finished.

CHAPTER 5

Lipschitz Mappings with Big Images

Our first step in the proof of local uniform rectifiability for quasiminimizers will be to find Lipschitz mappings from S to $\mathbf{R}^d$ with large images (in terms of measure). This is the goal of the present chapter. Later on we shall show that there are reasonably large subsets of S^* on which these mappings into $\mathbf{R}^d$ are bilipschitz, as in Definition 2.3.

PROPOSITION 5.1. *Let U be an open set in $\mathbf{R}^n$, let S be a (U, k, δ)-quasiminimizer for H^d, and let Q be a cube centered on S^* which satisfies*

(5.2) $$2Q \subseteq U \quad \text{and} \quad \operatorname{diam} Q < \delta.$$

Then there is a C-Lipschitz map $h : \mathbf{R}^n \to \mathbf{R}^d$ such that

(5.3) $$H^d\bigl(h(S^* \cap Q)\bigr) \geq C^{-1}(\operatorname{diam} Q)^d,$$

where C depends only on n and k.

To prove this, fix a cube Q as in the statement of Proposition 5.1, and let j be a positive integer, to be chosen later. Let Q_1 be a cube which is concentric with Q and satisfies

(5.4) $$\tfrac{1}{2}Q \subseteq Q_1 \subseteq Q$$

and

(5.5) $$H^d\bigl(S \cap A_j(Q_1)\bigr) \leq C 2^{-j} H^d(S \cap Q),$$

where $A_j(Q_1) = Q_1 \setminus \operatorname{int}[(1 - 2^{-j+1})Q_1]$. (This is the same as the union of the cubes $R \in \Delta_j(Q_1)$ that touch ∂Q_1, as in (4.6).) The existence of such a cube Q_1 follows easily from an averaging argument (i.e., the "average" choice of Q_1 which satisfies (5.4) will also satisfy (5.5)).

Notice that $E = S^* \cap Q_1$ is semi-regular of dimension d (Definition 3.29), and with a constant that depends only on k and n. This follows from Proposition 4.1, in the same way that Ahlfors regularity implies semi-regularity of the same dimension. (See the argument given just after Definition 3.29.)

Let $\phi : \mathbf{R}^n \to \mathbf{R}^n$ be the C-Lipschitz mapping obtained by applying Lemma 3.31 to Q_1 and $E = S^* \cap Q_1$. Set

(5.6) $$W = \{x \in \mathbf{R}^n : \phi(x) \neq x\},$$

as usual. Then

$$W \cup \phi(W) \subseteq Q_1 \subseteq Q,$$

by (3.2) and (3.5) (as provided by Lemma 3.31), and ϕ satisfies the usual requirements (1.3), (1.5), (1.6), and (1.7), by (5.2) and Remark 1.10. From (1.8) we get

that
$$(5.7) \qquad H^d(S \cap W) \leq k H^d\big(\phi(S \cap W)\big).$$
Put
$$(5.8) \qquad F = \phi\big(S \cap (Q_1 \backslash A_j(Q_1))\big).$$
We want to show that F is not too small (in terms of measure). Observe first that $\phi(S \cap W) \subseteq F \cup \phi(S \cap A_j(Q_1))$, since $W \subseteq Q_1$. Thus we have that
$$(5.9) \qquad \begin{aligned} H^d\big(\phi(S \cap W)\big) &\leq H^d(F) + H^d\big(\phi(S \cap A_j(Q_1))\big) \\ &\leq H^d(F) + C H^d\big(S \cap A_j(Q_1)\big) \\ &\leq H^d(F) + C' 2^{-j} H^d(S \cap Q) \\ &\leq H^d(F) + C'' 2^{-j} (\operatorname{diam} Q)^d, \end{aligned}$$
using the fact that ϕ is C-Lipschitz (by Lemma 3.31) for the second inequality, the estimate (5.5) for the third inequality, and Proposition 4.1 for the last step. All of the constants here depend only on n and k.

On the other hand, if $x \in [S \cap (Q_1 \backslash A_j(Q_1))]\backslash F$, then $\phi(x) \neq x$ by definition of F, and hence $x \in W$. This implies that
$$(5.10) \qquad \begin{aligned} H^d(S \cap W) &\geq H^d\big([S \cap (Q_1 \backslash A_j(Q_1))]\backslash F\big) \\ &\geq H^d\big(S \cap (Q_1 \backslash A_j(Q_1))\big) - H^d(F) \\ &\geq H^d\big(S \cap \tfrac{1}{4}Q\big) - H^d(F) \\ &\geq C^{-1}(\operatorname{diam} Q)^d - H^d(F). \end{aligned}$$
This uses (5.4) and the lower bound in Proposition 4.1, and we should restrict ourselves to $j > 2$ in order to ensure that $A_j(Q_1)$ does not intersect $\frac{1}{2}Q_1$. Combining this with (5.7) and (5.9) we obtain
$$(5.11) \qquad C^{-1}(\operatorname{diam} Q)^d - H^d(F) \leq k H^d(F) + C'' k 2^{-j}(\operatorname{diam} Q)^d.$$

We now choose j to be the smallest integer (greater than 2) such that
$$C'' k 2^{-j} < \frac{1}{2} C^{-1},$$
with C and C'' as in (5.11). This makes sense, because C and C'' do not depend on j. Note that j depends only on n and k, because of the corresponding features of C and C''. With this choice of j we have that
$$(5.12) \qquad H^d(F) \geq \frac{1}{2}(k+1)^{-1} C^{-1}(\operatorname{diam} Q)^d.$$

Observe that $\phi\big(S^* \cap (Q_1 \backslash A_j(Q_1))\big)$ does not intersect ∂Q_1, because of (3.5). Hence it is contained in $\mathcal{S}_{j,d}(Q_1)$, by (3.4). (Remember that ϕ is obtained by applying Lemma 3.31 to Q_1 with $E = S^* \cap Q_1$.) Since we also have that $H^d\big(\phi(S\backslash S^*)\big) = H^d(S\backslash S^*) = 0$, we get that
$$(5.13) \qquad H^d\big(F \backslash \mathcal{S}_{j,d}(Q_1)\big) = 0.$$
On the other hand, $\mathcal{S}_{j,d}(Q_1)$ consists of a bounded number of d-dimensional faces $T \in \Delta_{j,d}(Q_1)$, since we have a uniform bound for j. This implies that there must be a face T in $\Delta_{j,d}(Q_1)$ such that
$$(5.14) \qquad H^d(F \cap T) \geq C_1^{-1}(\operatorname{diam} Q)^d$$

for a suitable constant C_1 (that depends only on n and k). If $\pi : \mathbf{R}^n \to \mathbf{R}^d$ is a 1-Lipschitz linear mapping whose restriction to T is an isometry, then $h = \pi \circ \phi$ satisfies the requirements of Proposition 5.1, with (5.3) coming from (5.14). This completes the proof of Proposition 5.1.

REMARK 5.15. For various reasons, it would be nicer to show directly that S^* has large projections (in the sense of Definition 2.9), rather than the intermediate result in Proposition 5.1. This would amount to saying that the mapping h in Proposition 5.1 can be taken to be a projection, instead of a more complicated nonlinear mapping, as in the proof above. We shall derive the property of large projections later on, in Chapter 10, after establishing the uniform rectifiability of S^*, but a direct approach also seems to be feasible. This direct approach would give the lower bound in the Ahlfors-regularity condition (4.2), but unfortunately the argument is fairly complicated, and we shall not pursue it here.

CHAPTER 6

From Lipschitz Functions to Projections

In this chapter we shall give a construction which will permit us (in practice) to reduce to the situation where the mapping h from Proposition 5.1 is an orthogonal projection, through a suitable transformation of the quasiminimizing set S. This will be convenient for some of our subsequent arguments (in Chapters 8 and 9).

PROPOSITION 6.1. *Let U be an open set in $\mathbf{R}^n$, let S be a (U, k, δ)-quasiminimizer for H^d, and let $h : \mathbf{R}^n \to \mathbf{R}^\ell$ be an arbitrary Lipschitz mapping. Define $\widehat{U} \subseteq \mathbf{R}^n \times \mathbf{R}^\ell$ and $\widehat{S} \subseteq \widehat{U}$ by*

(6.2) $$\widehat{U} = \{(x, y) \in \mathbf{R}^n \times \mathbf{R}^\ell : x \in U\}$$

and

(6.3) $$\widehat{S} = \{(x, y) \in \mathbf{R}^n \times \mathbf{R}^\ell : x \in S \text{ and } y = h(x)\}.$$

Then $\widehat{S}$ is a $(\widehat{U}, \widehat{k}, \delta)$-quasiminimizer, where $\widehat{k}$ depends only on d, k, and the Lipschitz norm of h.

The proof of this amounts to a fairly straightforward verification of the definition of a quasiminimizer (using the corresponding properties for S). Let C_0 denote the Lipschitz norm of h. Clearly, $\widehat{S} \neq \emptyset$, because $S \neq \emptyset$. The closure of $\widehat{S}$ in $\mathbf{R}^{n+\ell}$ is $\{(x, y) : x \in \overline{S} \text{ and } y = h(x)\}$, and so $\widehat{S}$ is relatively closed in $\widehat{U}$ since S is relatively closed in U (by definition of a quasiminimizer). Let $\pi : \mathbf{R}^n \times \mathbf{R}^\ell \to \mathbf{R}^n$ denote the standard projection onto the first set of coordinates, i.e.,

$$\pi(x, y) = x.$$

If B is a ball contained in $\widehat{U}$, then the projection $\pi(B)$ of B in $\mathbf{R}^n$ is contained in U, and

$$H^d(\widehat{S} \cap B) \leq C H^d(S \cap \pi(B)) < +\infty,$$

because $\widehat{S} \cap B$ is contained in the graph of h over $\pi(B)$. Thus $\widehat{S}$ satisfies the first two conditions (1.1) and (1.2), and it remains to establish the comparison property (1.8).

Let $\widehat{\phi} : \mathbf{R}^n \times \mathbf{R}^\ell \to \mathbf{R}^n \times \mathbf{R}^\ell$ be a Lipschitz mapping. Put

(6.4) $$\widehat{W} = \{(x, y) \in \mathbf{R}^n \times \mathbf{R}^\ell : \widehat{\phi}(x, y) \neq (x, y)\},$$

as in (1.4), and assume that $\widehat{\phi}$ satisfies the analogues of (1.5), (1.6), and (1.7) for $\widehat{S}$ and $\widehat{U}$. We want to define a mapping $\phi : \mathbf{R}^n \to \mathbf{R}^n$, to which we can apply (1.8) in order to derive information about $\widehat{S}$ and $\widehat{\phi}$ from the quasiminimizing property of S. Set

(6.5) $$\phi(x) = \pi \circ \widehat{\phi}(x, h(x)),$$

where $\pi : \mathbf{R}^n \times \mathbf{R}^\ell \to \mathbf{R}^n$ is the obvious projection, as above. Clearly $\phi : \mathbf{R}^n \to \mathbf{R}^n$ is Lipschitz. Let $W = \{x \in \mathbf{R}^n : \phi(x) \neq x\}$. If $x \in W$, then $(x, h(x)) \in \widehat{W}$ (by definition of ϕ and $\widehat{W}$), and therefore x lies in $\pi(\widehat{W})$ and $\phi(x) = \pi(\widehat{\phi}(x, h(x)))$ lies in $\pi(\widehat{\phi}(\widehat{W}))$. Hence

$$(6.6) \qquad W \cup \phi(W) \subseteq \pi(\widehat{W} \cup \widehat{\phi}(\widehat{W})).$$

Because of this,

$$(6.7) \qquad \operatorname{diam}(W \cup \phi(W)) \leq \operatorname{diam}(\widehat{W} \cup \widehat{\phi}(\widehat{W})) < \delta,$$

by the analogue of (1.5) for $\widehat{\phi}$, and similarly

$$(6.8) \qquad \begin{aligned} \operatorname{dist}&(W \cup \phi(W), \mathbf{R}^n \backslash U) \\ &\geq \operatorname{dist}(\widehat{W} \cup \widehat{\phi}(\widehat{W}), (\mathbf{R}^n \backslash U) \times \mathbf{R}^\ell) \\ &= \operatorname{dist}(\widehat{W} \cup \widehat{\phi}(\widehat{W}), (\mathbf{R}^n \times \mathbf{R}^\ell) \backslash \widehat{U}) > 0 \end{aligned}$$

by definition of $\widehat{U}$ and the analogue of (1.6) for $\widehat{\phi}$. (The first inequality in (6.8) can be replaced by an equality, but we do not need that.)

Next we consider the homotopy condition (1.7). Let

$$\widehat{h} : [0,1] \times \mathbf{R}^n \times \mathbf{R}^\ell \to \mathbf{R}^n \times \mathbf{R}^\ell$$

be the homotopy given by (1.7) for $\widehat{\phi}$. Thus $\widehat{h}$ is continuous, $\widehat{h}(0, x, y) = (x, y)$, $\widehat{h}(1, x, y) = \widehat{\phi}(x, y)$, $\widehat{h}(t, \cdot)$ is Lipschitz for each $t \in [0,1]$, and $\widehat{h}$ has the following "support" property. Given $t \in [0,1]$, set $\widehat{\phi}_t(x,y) = \widehat{h}(t,x,y)$, and define $\widehat{W}_t$ for each $\widehat{\phi}_t$ as in (6.4). If $\widehat{\mathcal{H}}$ denotes the union of all the sets $\widehat{W}_t \cup \widehat{\phi}_t(\widehat{W}_t)$, $0 \leq t \leq 1$, then the support condition on the homotopy $\widehat{h}$ requires that $\operatorname{diam} \widehat{\mathcal{H}} < \delta$ and $\operatorname{dist}(\widehat{\mathcal{H}}, (\mathbf{R}^n \times \mathbf{R}^\ell) \backslash \widehat{U}) > 0$. (See (1.7b).)

For each $t \in [0,1]$, define $\phi_t : \mathbf{R}^n \to \mathbf{R}^n$ by $\phi_t(x) = \pi(\widehat{\phi}_t(x, h(x)))$. This is the same formula as (6.5), except that $\widehat{\phi}$ has been replaced by $\widehat{\phi}_t$. In particular, $\phi_1 = \phi$. Notice that $\phi_0(x) = \pi(\widehat{\phi}_0(x, h(x))) = \pi((x, h(x))) = x$. Thus, if we set $h(t, x) = \phi_t(x)$, then h is a (continuous) homotopy from x to $\phi(x)$. This homotopy is Lipschitz in x for each fixed t, because of the corresponding requirement for $\widehat{h}$. The support condition for h is also easy to check: if $W_t = \{x \in \mathbf{R}^n : \phi_t(x) \neq x\}$, then the same argument as for (6.6) shows that

$$(6.9) \qquad W_t \cup \phi_t(W_t) \subseteq \pi(\widehat{W}_t \cup \widehat{\phi}_t(\widehat{W}_t)).$$

Thus, if $\mathcal{H}$ denotes the union of all the sets $W_t \cup \phi_t(W_t)$, we get that $\mathcal{H} \subseteq \pi(\widehat{\mathcal{H}})$. Therefore, $\operatorname{diam} \mathcal{H} \leq \operatorname{diam} \widehat{\mathcal{H}} < \delta$ and

$$\begin{aligned} \operatorname{dist}(\mathcal{H}, \mathbf{R}^n \backslash U) &\geq \operatorname{dist}(\widehat{\mathcal{H}}, (\mathbf{R}^n \backslash U) \times \mathbf{R}^\ell) \\ &= \operatorname{dist}(\widehat{\mathcal{H}}, (\mathbf{R}^n \times \mathbf{R}^\ell) \backslash \widehat{U}) > 0, \end{aligned}$$

as required in (1.7). (Again, the inequality here is actually an equality, but we do not need that.)

This completes the verification that ϕ satisfies (1.3), (1.5), (1.6), and (1.7). We may now apply (1.8) to obtain that

$$(6.10) \qquad H^d(S \cap W) \leq k H^d(\phi(S \cap W)).$$

We want to use this inequality to control $H^d(\widehat{S} \cap \widehat{W})$. We begin by setting

(6.11) $\quad\quad \Sigma_1 = \{(x,y) \in \mathbf{R}^n \times \mathbf{R}^\ell : x \in S \cap W \text{ and } y = h(x)\}.$

We have already seen that $(x, h(x)) \in \widehat{W}$ when $x \in W$, and so $\Sigma_1 \subseteq \widehat{S} \cap \widehat{W}$. Using (6.10), we get that

(6.12) $\quad\quad H^d(\Sigma_1) \leq C H^d(S \cap W) \leq Ck H^d(\phi(S \cap W)),$

where C depends only on d and C_0 (the Lipschitz constant for h). From the definitions (6.5) and (6.11) of $\widehat{\phi}$ and Σ_1 we have that $\phi(S \cap W) = \pi(\widehat{\phi}(\Sigma_1))$, so that $H^d(\phi(S \cap W)) \leq H^d(\widehat{\phi}(\Sigma_1))$. Therefore

(6.13) $\quad\quad H^d(\Sigma_1) \leq Ck H^d(\widehat{\phi}(\Sigma_1)),$

by (6.12).

Now consider $\Sigma_2 = \widehat{S} \cap \widehat{W} \setminus \Sigma_1$. If $(x,y) \in \Sigma_2$, then $x \in S$ and $y = h(x)$ because $(x,y) \in \widehat{S}$. Since $(x,y) \notin \Sigma_1$, we have that $x \notin W$, and hence $\phi(x) = x$. This (and (6.5)) imply that

(6.14) $\quad\quad \pi \circ \widehat{\phi} = \pi \text{ on } \Sigma_2.$

On the other hand, $H^d(\Sigma_2) \leq C H^d(\pi(\Sigma_2))$, where C depends on d and C_0, because Σ_2 lies on the graph of h (since $\Sigma_2 \subseteq \widehat{S}$). We also have that $H^d(\pi(\Sigma_2)) = H^d(\pi(\widehat{\phi}(\Sigma_2)))$, by (6.14), and that $H^d(\pi(\widehat{\phi}(\Sigma_2))) \leq H^d(\widehat{\phi}(\Sigma_2))$ (automatically). Altogether,

(6.15) $\quad\quad H^d(\Sigma_2) \leq C H^d(\widehat{\phi}(\Sigma_2)).$

This combined with (6.13) yields

(6.16) $\quad\quad H^d(\widehat{S} \cap \widehat{W}) \leq Ck H^d(\widehat{\phi}(\widehat{S} \cap \widehat{W})),$

with a constant C that depends only on d and C_0. Thus $\widehat{S}$ satisfies the analogue of (1.8), with $\widehat{k} = Ck$. This completes the proof of Proposition 6.1.

CHAPTER 7

Regular Sets and Cubical Patchworks

So far we have been able to show that there are Lipschitz mappings h from S into $\mathbf{R}^d$ with "large images" (in the sense of Proposition 5.1), and we have given a recipe (in Proposition 6.1) which will permit us to operate in some situations as if such a mapping h were a projection. We still need to show that there are reasonably large subsets of S^* on which such mappings are bilipschitz. To this end we shall employ a stopping-time argument from [4]. This argument relies on the existence of decompositions of a given Ahlfors-regular set which are analogous to the usual dyadic decompositions of ordinary Euclidean space. In the present chapter we discuss the existence of similar decompositions for our set S^*, using (only) the local Ahlfors-regularity property given in Proposition 4.1.

Let us first be more precise about the kind of decompositions that we want to have.

DEFINITION 7.1. *Let $F \subseteq \mathbf{R}^n$ be a bounded Ahlfors-regular set of dimension d (as in Definition 2.1), and set $R = \operatorname{diam} F$. A* cubical patchwork *on F is a sequence $\{\Sigma_j\}_{j=0}^{\infty}$ of partitions of F into measurable subsets Q, $Q \in \Sigma_j$, which satisfies the following properties (for some constant C):*

(7.2)   if $Q \in \Sigma_j$, then $C^{-1} 2^{-j} R \leq \operatorname{diam} Q \leq C 2^{-j} R$ and
$$C^{-1}(2^{-j}R)^d \leq H^d(Q) \leq C(2^{-j}R)^d;$$

(7.3)   if $j \leq j'$, $Q \in \Sigma_j$, and $Q' \in \Sigma_{j'}$, then either
$$Q' \subseteq Q \text{ or } Q' \cap Q = \emptyset;$$

(7.4) $H^d(\{x \in Q : \operatorname{dist}(x, F \backslash Q) \leq \tau \operatorname{diam} Q\})$
$$+ H^d(\{x \in F \backslash Q : \operatorname{dist}(x, Q) \leq \tau \operatorname{diam} Q\}) \leq C \tau^{1/C} (\operatorname{diam} Q)^d$$
for all $Q \in \bigcup_j \Sigma_j$ and all $\tau \in (0, 1)$.

Notice that these conditions are all satisfied by the usual dyadic partitions of a cube in $\mathbf{R}^d$. In particular, the "small boundary" property (7.4) reflects some of the smoothness enjoyed by the boundaries of ordinary Euclidean cubes. We should perhaps emphasize that (7.4) is concerned only with the behavior of Q inside F, and not with the way that either of them sits inside $\mathbf{R}^n$.

PROPOSITION 7.5. *If F is a bounded Ahlfors-regular set of dimension d in $\mathbf{R}^n$, then F admits a cubical patchwork $\{\Sigma_j\}_{j \geq 0}$ as in Definition 7.1, with a constant C that depends only on n, d, and the Ahlfors-regularity constant for F.*

This follows from a construction in [4], although the formulation in Appendix I of [5] is closer to our present needs. There is a minor difference, however, in that the existence of the partitions $\{\Sigma_j\}$ is discussed and established in [4] and [5] for unbounded regular sets. This is not a real problem, for two reasons. The first is that the proofs given in [4] and [5] work equally well for bounded regular sets. (In fact, there was actually an extra argument in [4] and [5] to accommodate the unboundedness of the given set.) The second reason is that if F is a bounded regular set, then it is very easy to derive the existence of a cubical patchwork for F from the analogous result for unbounded sets. To see this, consider the set $\widehat{F} = F \cup P$, where P is a d-plane whose distance to F is equal to $R = \operatorname{diam} F$. This is an unbounded regular set, to which we can apply [4], [5] to obtain partitions $\widehat{\Sigma}_j$ of $\widehat{F}$ which satisfy analogues of (7.2), (7.3), and (7.4), but with j now running through $\mathbf{Z}$ and F replaced with $\widehat{F}$. (Since $\widehat{F}$ is unbounded, we should not use $R = \operatorname{diam} \widehat{F}$ in (7.2); for compatibility with [4] and [5], let us take $R = 1$ for $\{\widehat{\Sigma}_j\}$ and $\widehat{F}$.)

Because of the version of (7.2) for $\{\widehat{\Sigma}_j\}$ and $\widehat{F}$, there is a smallest integer n_0 such that if $\widehat{Q} \in \widehat{\Sigma}_j$ with $j \geq n_0$, then $\widehat{Q}$ is either wholly contained in F, or wholly contained in the d-plane P (and hence disjoint from F). By taking n_0 to be as small as possible we ensure that 2^{-n_0} is bounded from above and below by constant multiples of $\operatorname{diam} F$. (This uses the version of (7.2) for $\{\widehat{\Sigma}_j\}$ and $\widehat{F}$, and the way that we chose P.) We define Σ_j for $j \geq 0$ to be the collection of Q's in $\widehat{\Sigma}_{j+n_0}$ such that $Q \subseteq F$. It is not hard to see that $\{\Sigma_j\}_{j=0}^{\infty}$ satisfies (7.2)–(7.4), and with a constant C which is not too much different from the one for $\{\widehat{\Sigma}_j\}$. This gives Proposition 7.5.

PROPOSITION 7.6. *Let B be a closed ball in $\mathbf{R}^n$, and let E be a nonempty closed subset of $2B$. We assume that B is centered on E, and that*

(7.7) $$C_0^{-1} r^d \leq H^d(E \cap B(x,r)) \leq C_0 r^d$$

for some $C_0 > 0$ and all pairs (x,r) that satisfy

(7.8) $$x \in E \cap \left(\tfrac{3}{2}B\right) \quad \text{and} \quad 0 < r \leq \operatorname{diam} E.$$

Under these conditions, there is a compact Ahlfors-regular set F of dimension d in $\mathbf{R}^n$ such that

(7.9) $$E \cap B \subseteq F \subseteq E \cap \left(\tfrac{3}{2}B\right),$$

and a cubical patchwork $\{\Sigma_j\}_{j \geq 0}$ for F which is adapted to E in the sense that

(7.10) $$H^d(\{x \in Q : \operatorname{dist}(x, E \backslash Q) \leq \tau \operatorname{diam} Q\})$$
$$+ H^d(\{x \in E \backslash Q : \operatorname{dist}(x, Q) \leq \tau \operatorname{diam} Q\})$$
$$\leq C \tau^{1/C} (\operatorname{diam} Q)^d$$

for some constant C and all $Q \in \bigcup_j \Sigma_j$ and $\tau \in (0,1)$. (This is stronger than (7.4), because it uses E instead of F.) The Ahlfors-regularity constant for F and the constants in (7.10), (7.2), and (7.4) (for this choice of $\{\Sigma_j\}_{j \geq 0}$) are bounded in a way that depends only on C_0 and n.

This proposition will be employed in the next chapters, with $E = S^* \cap \tfrac{5}{3} B$, say, and where B is a ball of radius $< \delta$ that is centered on S^* and satisfies $2B \subseteq U$. In this case the hypothesis (7.7) of Proposition 7.6 will come from Proposition 4.1.

For the rest of this chapter, E and B will be as in the statement of Proposition 7.6. Notice that $H^d(E) \leq C_0 (\operatorname{diam} E)^d$, because of (7.7), and hence

(7.11) $$H^d(E \cap B(x,r)) \leq C r^d \quad \text{for all } x \in E \cap \tfrac{3}{2} B \text{ and } r > 0.$$

(In other words, we do not need to restrict the size of r for the upper estimate in (7.8).) We shall first construct a regular set F_0 which contains $E \cap B$ but is not necessarily contained in E, as in the next lemma.

LEMMA 7.12. *There is a compact Ahlfors-regular set F_0 of dimension d such that*

(7.13) $$F_0 \subseteq \tfrac{3}{2}B \quad and \quad F_0 \cap B_1 = E \cap B_1,$$

where $B_1 = \tfrac{11}{10}B$. The Ahlfors-regularity constant may be chosen in such a way as to depend only on the constant C_0 from (7.7) and the dimension n.

To prove the lemma, let $\{Q_i\}_{i \in I}$ be a Whitney decomposition of $\mathbf{R}^n \setminus B_1$, in the sense of Section IV.1 in [29]. Specifically, the Q_i's should be (standard) cubes in $\mathbf{R}^n$ that are contained in $\mathbf{R}^n \setminus B_1$, have disjoint interiors, cover $\mathbf{R}^n \setminus B_1$, and satisfy

(7.14) $$\frac{1}{3} \operatorname{dist}(Q_i, B_1) \leq \operatorname{diam} Q_i \leq \operatorname{dist}(Q_i, B_1)$$

for all $i \in I$. The latter is slightly different from the requirements in [29], and it can easily be accommodated by subdividing the cubes given in [29].

Set

(7.15) $$J = \{i \in I : Q_i \subseteq \tfrac{3}{2}B \quad \text{and} \quad Q_i \cap E \neq \emptyset\}$$

and, for each $x \in B_1$ and $r > 0$,

(7.16) $$J(x, r) = \{i \in J : Q_i \cap B(x, r) \neq \emptyset\}.$$

Let us verify that

(7.17) $$\sum_{i \in J(x,r)} (\operatorname{diam} Q_i)^d \leq C r^d$$

for all $x \in B_1$ and $r > 0$.

Notice first that $H^d(E \cap \tfrac{11}{10} Q_i) \geq C^{-1}(\operatorname{diam} Q_i)^d$ for $i \in J$, because of our regularity assumption (7.7). Using (7.14), it is not hard to check that the cubes $\tfrac{11}{10}Q_i$, $i \in I$, have bounded overlap. This implies that

(7.18) $$\sum_{i \in J(x,r)} (\operatorname{diam} Q_i)^d \leq C \sum_{i \in J(x,r)} H^d(E \cap \tfrac{11}{10}Q_i)$$
$$\leq C' H^d\Big(E \cap \Big[\bigcup_{i \in J(x,r)} \tfrac{11}{10}Q_i\Big]\Big),$$

for a suitable constant C'. If $i \in J(x,r)$, then $\operatorname{diam} Q_i \leq \operatorname{dist}(Q_i, B_1) \leq r$, because of (7.14) and the requirements that x lie in B_1 and that Q_i intersect $B(x,r)$. Therefore

$$\bigcup_{i \in J(x,r)} \tfrac{11}{10}Q_i \subseteq B(x, 3r),$$

and the right hand side of (7.18) is at most $C' H^d(E \cap B(x, 3r))$. We may as well assume that $J(x,r) \neq \emptyset$ – since otherwise (7.17) is trivial – and this ensures that $E \cap \big(\tfrac{3}{2}B\big) \cap B(x, 3r)$ is not empty, by the definition (7.16) of $J(x,r)$. From here we obtain that

$$H^d(E \cap B(x, 3r)) \leq C r^d,$$

because of (7.11). (Note that we may not apply (7.11) with the same x as we have here, unless x happens to lie in E. In general, we can use (7.11) with x replaced by an element of $E \cap B(x, 3r)$, and with $6r$ instead of r. This is adequate for the inequality above, but with a slightly different constant C from the one in (7.11).) This proves (7.17), because of (7.18).

For each $i \in J$, let D_i denote the intersection of Q_i with a d-plane which passes through the center of Q_i. The specific choice of d-plane does not matter. Define F_0 by

$$(7.19) \qquad F_0 = (E \cap B_1) \cup \left(\bigcup_{i \in J} D_i \right).$$

Observe that $F_0 \subseteq \frac{3}{2}B$, since each Q_i, $i \in J$, is contained in $\frac{3}{2}B$ (by (7.15)). We also have that $F_0 \cap B_1 = E \cap B_1$, because $D_i \subseteq Q_i \subseteq \mathbf{R}^n \backslash B_1$ for each $i \in J$. This proves (7.13), and we want to show that F_0 is Ahlfors regular.

Let us first check that F_0 is closed. Let $\{z_j\}$ be any convergent sequence in F_0, and let z denote its limit. If infinitely many z_j's lie in $E \cap B_1$, or in a single D_i, then $z \in F_0$. If neither of these possibilities hold, then each z_j lies in some $D_{i(j)}$ for j large enough, and the diameters of the corresponding cube $Q_{i(j)}$ necessarily tends to zero as $j \to \infty$. (That is, if the diameters of the $Q_{i(j)}$'s did not tend to zero, then infinitely many of the $Q_{i(j)}$'s would have to come from a finite collection of Q_i's (by the geometry of the Whitney cubes), and we would be back in the first case.) In this situation we have that $z \in E \cap B_1$, because of (7.14), (7.15), and the assumption in Proposition 7.6 that E be closed. This proves that F_0 is closed.

Next we want to check that

$$(7.20) \qquad H^d\big(F_0 \cap B(x, r)\big) \leq Cr^d \quad \text{for all} \quad x \in \mathbf{R}^n \quad \text{and} \quad r > 0.$$

We already have that

$$(7.21) \qquad H^d\big(E \cap B_1 \cap B(x, r)\big) \leq Cr^d \quad \text{for} \quad x \in \mathbf{R}^n \quad \text{and} \quad r > 0$$

by (7.7) or (7.11), and so it is enough to show that

$$(7.22) \qquad H^d\left(B(x, r) \cap \left[\bigcup_{i \in J} D_i\right]\right) \leq Cr^d$$

for all $x \in \mathbf{R}^n$ and $r > 0$.

When $x \in B_1$, this follows from (7.17) and (7.16). If $B(x, 2r) \cap B_1 \neq \emptyset$, then it is easy to reduce to the case where $x \in B$, by increasing slightly the choice of r. If $B(x, 2r)$ does not intersect B_1, then (7.14) ensures that $B(x, r)$ cannot intersect more than a bounded number of Q_i's, and (7.22) is immediate. Thus (7.22) holds in all cases, and (7.20) follows.

We are left with the task of establishing the lower bound

$$(7.23) \qquad H^d\big(F_0 \cap B(x, r)\big) \geq C^{-1} r^d$$

for $x \in F_0$ and $0 < r < \operatorname{diam} F_0$.

If $x \in D_i$ for some $i \in J$ such that $r \leq 100 \operatorname{diam} Q_i$, then

$$(7.24) \qquad H^d\big(F_0 \cap B(x, r)\big) \geq H^d\big(D_i \cap B(x, r)\big) \geq C^{-1} r^d$$

(by the definition of D_i), and we have what we want.

Assume now that $x \in D_i$ for some $i \in J$, but $r > 100 \operatorname{diam} Q_i$. Choose a point $y \in Q_i \cap E$. (This is possible, because of the definition (7.15) of J.) Let us first

7. REGULAR SETS AND CUBICAL PATCHWORKS

check that

(7.25) $$Q_j \subseteq B\left(x, \tfrac{3r}{10}\right) \text{ for every Whitney cube } Q_j$$
$$\text{that intersects } B\left(y, \tfrac{r}{10}\right).$$

Given such a cube Q_j, we have that

(7.26) $$\operatorname{dist}(Q_j, x) \leq \operatorname{dist}(Q_j, y) + |y - x|$$
$$\leq \frac{r}{10} + \operatorname{diam} Q_i,$$

since Q_j intersects $B\left(y, \tfrac{r}{10}\right)$ and both x, y lie in Q_i. We also have that

(7.27) $$\operatorname{diam} Q_j \leq \operatorname{dist}(Q_j, B_1) \leq \operatorname{dist}(Q_j, y) + \operatorname{dist}(y, B_1)$$
$$\leq \frac{r}{10} + \operatorname{dist}(y, B_1) \leq \frac{r}{10} + \operatorname{diam} Q_i + \operatorname{dist}(Q_i, B_1)$$
$$\leq \frac{r}{10} + 4 \operatorname{diam} Q_i.$$

This uses (7.14) for the first inequality, the assumption that Q_j intersects $B\left(y, \tfrac{r}{10}\right)$ for the third, the fact that y lies in Q_i in the next step, and then (7.14) again at the end. Combining (7.26) and (7.27) with the (assumed) inequality $\operatorname{diam} Q_i < \tfrac{r}{100}$ gives $Q_j \subseteq B\left(x, \tfrac{3r}{10}\right)$, which is what we wanted. This proves (7.25).

Using (7.25) we get that

(7.28) $$\bigcup_{j \in J\left(y, \tfrac{r}{10}\right)} Q_j \subseteq B\left(x, \tfrac{3r}{10}\right),$$

where $J\left(y, \tfrac{r}{10}\right) = \{j \in J : Q_j \cap B\left(y, \tfrac{r}{10}\right) \neq \emptyset\}$, as in (7.16) (even though $y \notin B_1$ here). Let us assume for the moment that

(7.29) $$r \leq r_0 = \operatorname{radius}(B)$$

and verify that

(7.30) $$(E \backslash B_1) \cap B\left(y, \tfrac{r}{10}\right) \subseteq \bigcup_{j \in J\left(y, \tfrac{r}{10}\right)} Q_j.$$

Since the cubes Q_j, $j \in I$, cover $\mathbf{R}^n \backslash B_1$, it suffices to show that $j \in J\left(y, \tfrac{r}{10}\right)$ whenever Q_j intersects $(E \backslash B_1) \cap B\left(y, \tfrac{r}{10}\right)$. Since Q_j intersects E and $B\left(y, \tfrac{r}{10}\right)$, it is enough to check that $Q_j \subseteq \tfrac{3}{2} B$ in order to conclude that $j \in J\left(y, \tfrac{r}{10}\right)$. (Compare with the definitions of $J\left(y, \tfrac{r}{10}\right)$ and J.) Since $x \in D_i$ ($\subseteq Q_i$) and $\operatorname{diam} Q_i < \tfrac{r}{100}$ in the case currently under consideration, we have that

$$\operatorname{dist}(x, B_1) \leq \operatorname{dist}(Q_i, B_1) + \operatorname{diam} Q_i \leq 4 \operatorname{diam} Q_i \leq \frac{4r}{100} \leq \frac{4r_0}{100},$$

by (7.14) and (7.29). From (7.25) we have that $Q_j \subseteq B\left(x, \tfrac{3r}{10}\right) \subseteq B\left(x, \tfrac{3r_0}{10}\right)$, and this implies that $Q_j \subseteq \tfrac{3}{2} B$, because of the definition of r_0 as the radius of B and the preceding bound for the distance from x to B_1. This proves that $j \in J\left(y, \tfrac{r}{10}\right)$, and (7.30) follows.

We are almost ready to prove (7.23) in the present case. We first use (7.30) and (7.11) to obtain that

$$H^d\big((E\backslash B_1) \cap B(y, \tfrac{r}{10})\big) \leq \sum_{j \in J(y, \frac{r}{10})} H^d(E \cap Q_j)$$

$$\leq C_0 \sum_{j \in J(y, \frac{r}{10})} \mathrm{diam}(Q_j)^d.$$

For the application of (7.11) in the second inequality, keep in mind that Q_j intersects $E \cap \frac{3}{2}B$ when $j \in J(y, \frac{r}{10}) \subseteq J$, by the definition (7.15) of J. (In other words, there is a point $x_j \in Q_j \cap E \cap \frac{3}{2}B$ to which (7.11) can be applied in order to get the inequality above.) From here we may conclude that

(7.31) $$H^d\big((E\backslash B_1) \cap B(y, \tfrac{r}{10})\big) \leq C \sum_{j \in J(y, \frac{r}{10})} H^d(D_j)$$

$$\leq C H^d\Big(B(x, r) \cap \Big[\bigcup_{j \in J} D_j\Big]\Big).$$

Specifically, the first inequality simply uses the fact that $H^d(D_j)$ is bounded from below by a constant multiple of $\mathrm{diam}(Q_j)^d$, by the definition of D_j (as the intersection of Q_j with a d-plane that passes through the center of Q_j; see the lines just before (7.19)). The second inequality in (7.31) relies on disjointness of the interiors of the Q_j's, to say that

$$\sum_{j \in J(y, \frac{r}{10})} H^d(D_j) = H^d\Big(\bigcup_{j \in J(y, \frac{r}{10})} D_j\Big).$$

We can do this because $H^d(D_j \cap \partial Q_j) = 0$ for every j, by the definition of D_j. We are also using (for the second inequality in (7.31)) the fact that $D_j \subseteq Q_j \subseteq B(x, r)$ when $j \in J(y, \frac{r}{10})$, as in (7.28).

As a complement to (7.31), let us check that

$$E \cap B_1 \cap B(y, \frac{r}{10}) \subseteq F_0 \cap B(x, r).$$

We automatically have that $F_0 \supseteq E \cap B_1$, by the definition (7.19) of F_0. We also have that $B(y, \frac{r}{10}) \subseteq B(x, r)$, because x and y both lie in Q_i and $\mathrm{diam}\, Q_i < \frac{r}{100}$. (This last comes from the properties of x and y that were "declared" just before (7.25).) Thus we get the inclusion displayed above.

Using this and (7.31) we conclude that

(7.32) $$H^d(E \cap B(y, \frac{r}{10})) \leq C H^d(F_0 \cap B(x, r)).$$

We want to apply (7.7) now to get that

$$H^d(E \cap B(y, \frac{r}{10})) \geq C_0^{-1} \big(\frac{r}{10}\big)^d.$$

In order to apply (7.7) in this way, we should check that $y \in \left(\frac{3}{2}B\right) \cap E$ and $\frac{r}{10} \leq \mathrm{diam}\, E$, as in (7.8). For the former, remember that $y \in Q_i \cap E$ and $i \in J$, as in the lines just before (7.25). Thus $Q_i \subseteq \frac{3}{2}B$, by the definition (7.15) of J, and $y \in \left(\frac{3}{2}B\right) \cap E$ follows. As for the bound for r, the assumption (7.29) that

$r \leq \text{radius}(B)$ ensures that $r \leq \text{diam}\, E$ (which is more than we need), since both the center of B and $x \in D_i \subseteq \mathbf{R}^n \backslash B$ lie in E by assumption.

Thus we may apply (7.7) in this way, and (7.32) yields

$$H^d\big(F_0 \cap B(x,r)\big) \geq (CC_0)^{-1}\Big(\frac{r}{10}\Big)^d.$$

This proves (7.23) in the present case where $x \in D_i$, $r > 100\,\text{diam}\, Q_i$, and $r \leq \text{radius}(B)$. (These hypotheses were made just before (7.25), and in (7.29).)

If $r > r_0 = \text{radius}(B)$ (and $x \in D_i$, $r > 100\,\text{diam}\, Q_i$ are still true), then we simply use the estimate

$$H^d\big(F_0 \cap B(x,r)\big) \geq H^d\big(F_0 \cap B(x,r_0)\big) \geq C^{-1} r_0^d,$$

where the first inequality is trivial and the second comes from the previous case. This is sufficient for (7.23), since we restrict ourselves to $r < \text{diam}\, F_0$ in (7.23), and because $\text{diam}\, F_0 \leq 3r_0$. (Remember that $F_0 \leq \frac{3}{2}B$, by construction. See (7.19) and the comments immediately after it.)

This shows that (7.23) holds when $x \in D_i$ and $r > 100\,\text{diam}\, Q_i$, whether or not $r \leq \text{radius}(B)$. We already covered the case where $x \in D_i$ and $r \leq 100\,\text{diam}\, Q_i$ in (7.24), and so we have now established (7.23) whenever $x \in D_i$ for some $i \in J$ and $0 < r < \text{diam}\, F_0$.

We are left with the case where $x \in E \cap B_1$. We may as well assume that $r \leq r_0$, because the estimate for $r_0 < r < \text{diam}\, F_0$ can then be derived from the estimate for $r = r_0$, for the same reasons as in the earlier case above.

The argument is similar to the previous one, but simpler. If Q_j is a Whitney cube that intersects $B\big(x, \frac{r}{10}\big)$, then

$$(7.33) \qquad \text{diam}\, Q_j \leq \text{dist}(Q_j, B_1) \leq \text{dist}(Q_j, x) \leq \frac{r}{10},$$

by (7.14) (and the fact that $x \in B_1$). In particular,

$$Q_j \subseteq B\Big(x, \frac{r}{5}\Big) \subseteq \frac{3}{2} B.$$

This uses $Q_j \cap B\big(x, \frac{r}{10}\big) \neq \emptyset$ and (7.33) for the first inclusion, and $r \leq r_0$, $x \in B_1$ for the second. From here we obtain that

$$(7.34) \qquad \bigcup_{j \in J(x, \frac{r}{10})} Q_j \subseteq B\Big(x, \frac{r}{5}\Big),$$

since $Q_j \cap B\big(x, \frac{r}{10}\big) \neq \emptyset$ when $j \in J\big(x, \frac{r}{10}\big)$ (as in the definition (7.16) of $J\big(x, \frac{r}{10}\big)$). We also have that

$$(7.35) \qquad (E \backslash B_1) \cap B\Big(x, \frac{r}{10}\Big) \subseteq \bigcup_{j \in J(x, \frac{r}{10})} Q_j.$$

Indeed, the Whitney cubes Q_j, $j \in I$, cover $\mathbf{R}^n \backslash B_1$, and so we only need to know that j has to lie in $J(x, r)$ when Q_j intersects $(E \backslash B_1) \cap B\big(x, \frac{r}{10}\big)$ in order to derive (7.35). This assertion is not hard to verify, using the definitions (7.16), (7.15) of $J\big(x, \frac{r}{10}\big)$, J, and the observation from above that $Q_j \subseteq \frac{3}{2}B$ when Q_j intersects $B\big(x, \frac{r}{10}\big)$.

At this point we can make the same computations as lead up to (7.31) to conclude that

$$H^d\Big((E\backslash B_1)\cap B\big(x,\tfrac{r}{10}\big)\Big) \le \sum_{j\in J(x,\frac{r}{10})} H^d(E\cap Q_j) \tag{7.36}$$

$$\le C_0 \sum_{j\in J(x,\frac{r}{10})} (\operatorname{diam} Q_j)^d$$

$$\le C \sum_{j\in J(x,\frac{r}{10})} H^d(D_j)$$

$$\le C H^d\Big(B(x,r)\cap \Big[\bigcup_{j\in J} D_j\Big]\Big).$$

In particular, the second inequality uses (7.11), and is justified in the same way as before, i.e., Q_j intersects $E \cap \left(\tfrac{3}{2}B\right)$ when $j \in J$, and hence when $j \in J(x,\tfrac{r}{10})$. (This ensures that there is a point $x_j \in Q_j$ to which (7.11) can be applied.) Adding $H^d(E\cap B_1)$ to both sides of the inequality we obtain that

$$H^d\Big(E\cap B\big(x,\tfrac{r}{10}\big)\Big) \le C H^d\big(F_0 \cap B(x,r)\big). \tag{7.37}$$

Next we want to apply (7.7) to get

$$H^d\Big(E\cap B\big(x,\tfrac{r}{10}\big)\Big) \ge C_0^{-1}\big(\tfrac{r}{10}\big)^d,$$

which would then imply (7.23) in this case, in combination with (7.37). To do this, we should check that

$$\tfrac{r}{10} < \operatorname{diam} E,$$

as in (7.8). Remember that we are assuming that $r \le r_0 = \operatorname{radius}(B)$, and that we only care about r's with $r < \operatorname{diam} F_0$ for (7.23). If we did have $\tfrac{r}{10} \ge \operatorname{diam} E$, then we would get $\operatorname{diam} E \le \tfrac{r_0}{10}$, and therefore $E \subseteq \tfrac{1}{10}B$ (since B is centered on E, as part of our original assumptions in Proposition 7.6). The definition (7.19) of F_0 would then collapse to $F_0 = E$, because J would be empty. (See (7.15), and remember that the Q_i's are contained in the complement of B_1.) In particular, we would have $\operatorname{diam} E = \operatorname{diam} F_0$. This would contradict the assumption that $\operatorname{diam} E \le \tfrac{r}{10}$, since we are only considering r's with $r < \operatorname{diam} F_0$. (For that matter, if F_0 is equal to E, then we do not need to deal with any of this anyway, in that the Ahlfors-regularity conditions for F_0 would follow immediately from their counterparts in (7.7) for E.)

Thus we have proved (7.23) in all cases, and we conclude that F_0 is Ahlfors regular of dimension d. This completes the proof of Lemma 7.12.

Let us return now to the proof of Proposition 7.6. Let B and E be as in Proposition 7.6, let F_0 be as in Lemma 7.12, and let $\{\Sigma_j\}_{j\ge 0}$ be a cubical patchwork for F_0 (whose existence is guaranteed by Proposition 7.5). Choose $j_0 \ge 0$ as small as possible so that every $Q \in \Sigma_{j_0}$ has diameter $\le \tfrac{r_0}{100}$, where r_0 (still) denotes the radius of B. Set

$$\Sigma_{j_0}^* = \{Q \in \Sigma_{j_0} : Q \cap B \ne \emptyset\}, \tag{7.38}$$

$$F_1 = \bigcup_{Q\in\Sigma_{j_0}^*} Q, \tag{7.39}$$

and $F = \overline{F}_1$. Notice that
$$E \cap B = F \cap B \subseteq F_1$$
and
$$F \subseteq F_0 \cap \left(\frac{101}{100}B\right) \subseteq E \cap \left(\frac{3}{2}B\right),$$
by (7.13) and the definition of j_0. Thus (7.9) holds. Let us try to define a cubical patchwork for F, and deal with the Ahlfors regularity of F afterwards.

We shall ignore the indices $j < j_0$ for the moment. For $j \geq j_0$, set

(7.40) $$\Sigma_j^* = \{Q \in \Sigma_j : Q \subseteq F_1\}.$$

These almost provide the decompositions of F that we want, but they do not take the elements of $F \backslash F_1$ into account. We shall see later that $H^d(F \backslash F_1) = 0$, but we still have to be a little careful in the way that the points in $F \backslash F_1$ are distributed among the Q's in Σ_j^*.

To do this, we begin by choosing any linear ordering for the set of cubes in $\Sigma_{j_0}^*$. Next we choose a linear order on $\Sigma_{j_0+1}^*$ which is compatible with the one on $\Sigma_{j_0}^*$, in the following sense. Let Q_1 and Q_2 be elements of $\Sigma_{j_0+1}^*$ such that $Q_1 \prec Q_2$ (with respect to the ordering on $\Sigma_{j_0+1}^*$). Suppose that Q_1', Q_2' are the elements of $\Sigma_{j_0}^*$ that contain Q_1 and Q_2, respectively. (Q_1' and Q_2' exist and are unique, because the Σ_j's satisfy (7.3).) We require that either $Q_1' = Q_2'$ or that $Q_1' \prec Q_2'$ in $\Sigma_{j_0}^*$. This is easy to arrange. We can continue this process indefinitely, to get a linear ordering on each Σ_j^*, $j > j_0$, which is compatible with the previous orderings on Σ_ℓ^* for $j_0 \leq \ell < j$ in the same sense as above. These orderings are like lexicographic orderings.

We should be a bit precise and say that the symbol $\prec$ denotes the "strict" version of the ordering, so that $Q_1 \prec Q_2$ implies that $Q_1 \neq Q_2$.

For each $j \geq j_0$ and each $Q \in \Sigma_j^*$, set

(7.41) $$\partial(Q) = [\overline{Q} \cap (F \backslash F_1)] \backslash \bigcup_{R \prec Q} \overline{R},$$

where the union is taken over all cubes $R \in \Sigma_j^*$ such that $R \prec Q$ with respect to the ordering discussed above. Roughly speaking, $\partial(Q)$ is the set of points in $F \backslash F_1$ that we want to add to Q in order to define a cubical patchwork for F. Thus we set

(7.42) $$\widehat{Q} = Q \cup \partial(Q),$$

and we want to show that the collections $\{\widehat{Q} : Q \in \Sigma_j^*\}$, $j \geq j_0$, are partitions of F with the required properties.

For each $j \geq j_0$, we know that the Q's from Σ_j^* form a partition of F_1, since Σ_j is a partition of F_0, and since (7.3) (together with (7.39)) tells us that every Q in Σ_j that intersects F_1 must be contained in F_1. In particular, F_1 is the union of the Q's in Σ_j^*, and $F = \overline{F}_1$ is the union of the closures of the Q's in Σ_j^*. Using (7.41) it is easy to check that the sets $\partial(Q)$, $Q \in \Sigma_j^*$, form a partition of $F \backslash F_1$, and that the sets $\widehat{Q}$, $Q \in \Sigma_j^*$, form a partition of F.

Let us verify the compatibility condition (7.3) for the $\widehat{Q}$'s. The main point is that if $Q \in \Sigma_j^*$, then

(7.43) $$\partial(Q) = \bigcup_{Q'} \partial(Q'),$$

where the union is taken over the children of Q, and "children" means the elements of Σ_{j+1}^* that are contained in Q. To see that (7.43) holds, notice first that $\overline{Q} = \bigcup_{Q'} \overline{Q'}$, where again the union is taken over the children Q' of Q in Σ_{j+1}^*. This equality holds because $Q = \bigcup_{Q'} Q'$, and because there are only finitely many children of Q. If Q_0' denotes the child of Q which is minimal for the ordering $\prec$, then

$$\bigcup_{R \prec Q} \overline{R} = \bigcup_{R' \prec Q_0'} \overline{R'},$$

where the R's on the left run through Σ_j^* and the R''s on the right run through Σ_{j+1}^*. This is easy to see from the compatibility property of the ordering: if $R \in \Sigma_j^*$ satisfies $R \prec Q$, then each child R' of R in Σ_{j+1}^* satisfies $R' \prec Q_0'$, and conversely if $R' \in \Sigma_{j+1}^*$ satisfies $R' \prec Q_0'$, then the parent $R \in \Sigma_j^*$ of R' satisfies $R \prec Q$. This last uses the minimality assumption on Q_0'. Once we have these observations, it is not hard to verify (7.43) using also the definition (7.41) of $\partial(Q)$. (In other words, the equality displayed above helps to account for everything that happens "before" Q and its children, "before" in the sense of our orderings on Σ_j^* and Σ_{j+1}^*, and then one is just left with what happens more directly among the children of Q.)

With the help of (7.43), one can check that the analogue of (7.3) holds for the $\widehat{Q}$'s when $j' = j+1$. The general case follows from repeated application of this one.

Let us turn now to the mass conditions. Notice first that

(7.44) $$H^d(F \backslash F_1) = 0.$$

Indeed, by its definition (7.39), F_1 is the union of the cubes Q in $\Sigma_{j_0}^*$, of which there are finitely many. Thus $F \backslash F_1$ is contained in the finite union of the sets

$$\overline{Q} \backslash Q = \{x \in F_0 \backslash Q : \operatorname{dist}(x, Q) = 0\}, \quad Q \in \Sigma_{j_0}^*.$$

Each of these sets has zero H^d-measure, by the small boundary condition (7.4) for the cubical patchwork $\{\Sigma_j\}_j$.

Thus $H^d(\partial(Q)) = 0$ and $H^d(\widehat{Q}) = H^d(Q)$ for all Q. We also have that $\operatorname{diam} \widehat{Q} = \operatorname{diam} Q$ for every $Q \in \Sigma_j$, since $Q \subseteq \widehat{Q} \subseteq \overline{Q}$, by construction. Thus the property (7.2) for the $\widehat{Q}$'s follows at once from the corresponding feature of the Q's.

Notice also that $\operatorname{dist}(x, \widehat{Q}) = \operatorname{dist}(x, Q)$ for all x and Q, again because $Q \subseteq \widehat{Q} \subseteq \overline{Q}$. Using this and the fact that $H^d(\widehat{Q} \backslash Q) = 0$, we obtain

$$H^d(\{x \in F_0 \backslash \widehat{Q} : \operatorname{dist}(x, \widehat{Q}) \leq \tau \operatorname{diam} \widehat{Q}\})$$
$$= H^d(\{x \in F_0 \backslash Q : \operatorname{dist}(x, Q) \leq \tau \operatorname{diam} Q\})$$

and hence

(7.45) $$H^d(\{x \in F_0 \backslash \widehat{Q} : \operatorname{dist}(x, \widehat{Q}) \leq \tau \operatorname{diam} \widehat{Q}\}) \leq C \tau^{1/C} (\operatorname{diam} Q)^d$$

for all $Q \in \Sigma_j^*$ and all $\tau \in (0,1)$, because of (7.4) for $\{\Sigma_j\}_j$. Similarly,
$$\operatorname{dist}(x, F_0 \backslash \widehat{Q}) \geq \operatorname{dist}(x, F_0 \backslash Q),$$
since $Q \subseteq \widehat{Q}$, and therefore
(7.46) $\qquad H^d(\{x \in \widehat{Q} : \operatorname{dist}(x, F_0 \backslash \widehat{Q}) \leq \tau \operatorname{diam} \widehat{Q}\})$
$$\leq H^d(\{x \in Q : \operatorname{dist}(x, F_0 \backslash Q) \leq \tau \operatorname{diam} Q\})$$
$$\leq C \tau^{1/C} (\operatorname{diam} Q)^d,$$
because of (7.4) for $\{\Sigma_j\}_j$ (and the fact that $H^d(\widehat{Q} \backslash Q) = 0$).

We now claim that the set F and the partitions $\{\widehat{Q} : Q \in \Sigma_j^*\}$ of F, $j \geq j_0$, satisfy the requirements of Proposition 7.6. (That is, $\{\widehat{Q} : Q \in \Sigma_j^*\}$ should be used in the role of Σ_j in the statement of Proposition 7.6.) Specifically, F is closed and satisfies (7.9) by construction. This was pointed out before, just after (7.39). The inequalities (7.45) and (7.46) contain the same information as in (7.10) (with Q replaced by $\widehat{Q}$). This is because E and F_0 coincide in $B_1 = \frac{11}{10}B$, while our cubical sets Q and $\widehat{Q}$ are contained in $\frac{101}{100}B$ for $Q \in \Sigma_j^*$, $j \geq j_0$, by construction.

In particular, the condition (7.10) (with Q replaced by $\widehat{Q}$) implies the analogue of (7.4) for the sets $\widehat{Q}$ in F, because $E \supseteq F$. We have already seen that the sets $\widehat{Q}$, $Q \in \Sigma_j^*$, satisfy the analogues of (7.3) and (7.2) as well (as discussed just before and after (7.44)).

It is not hard to see that F has to be an Ahlfors-regular set of dimension d. Indeed, the upper bound in (2.2) for F holds because $F \subseteq F_0$ and F_0 is Ahlfors regular of dimension d (as in Lemma 7.12). The lower bound in (2.2) can be derived from (7.2) for the sets $\widehat{Q}$ or Q, $Q \in \Sigma_j$, as follows. Fix a point $x \in F$ and a radius $r \leq \operatorname{diam} F$. From the construction of F, it is easy to see that there is a cube Q in some Σ_j^* such that $x \in \overline{Q}$ and $Q \subseteq B(x,r)$. (This also use the fact that the Σ_j's are partitions, and (7.3).) By choosing Q to be as large as possible, we can get that $\operatorname{diam} Q \geq C^{-1} r$ (for a suitable constant C). The lower bound for $H^d(B(x,r) \cap F)$ now follows from the one for $H^d(Q)$ in (7.2), since $B(x,r) \cap F \supseteq Q$.

There remains one small technical problem, which is that our partitions $\{\widehat{Q} : Q \in \Sigma_j^*\}$ of F are available only for $j \geq j_0$. A simple way to deal with this is to use the partition $\{\widehat{Q} : Q \in \Sigma_{j_0}^*\}$ of F also for the partitions associated to j with $0 \leq j < j_0$. This is slightly crude, but we do not lose control on the constants, because we know that j_0 cannot be too large. (The definition of j_0 was given just before (7.38).)

In summary, we have found an Ahlfors-regular set F and a cubical patchwork on F which satisfy the properties required in Proposition 7.6. It is not hard to see that the various constants involved can be bounded in a way that depends only on the constant C_0 from (7.7) and the dimension n, using the corresponding assertions for Proposition 7.5 and Lemma 7.12. This completes the proof of Proposition 7.6.

CHAPTER 8

A Stopping-Time Argument

In order to establish Theorem 2.11, about the uniform rectifiability of S^*, we need to come to grips with the following issue: given a Lipschitz mapping $h : S^* \to \mathbf{R}^d$ with large image, under what conditions can we find a substantial subset of S^* on which h is bilipschitz, and with uniform bounds? For this purpose we shall employ the method of [4]. Our first task will be to state the main lemma that contains the geometric information about S^* that we shall need. This lemma will be proved in Chapter 9. In the remainder of this chapter we shall explain how the local uniform rectifiability of S^* can be derived from this main lemma.

We begin by setting some notation which is relevant for the statement of the main lemma. Let S be a d-dimensional (U, k, δ) quasiminimizer, and let $B = \overline{B}(x_0, r_0)$ be a closed ball centered on S^*. We assume that

(8.1) $$r_0 < \delta \quad \text{and} \quad 2B \subseteq U.$$

Because of Proposition 4.1, $E = S^* \cap \left(\frac{5}{3}B\right) \subseteq 2B$ satisfies the hypotheses of Proposition 7.6 (with this choice of B). (More precisely, one applies Proposition 4.1 to small cubes centered around E, to get the necessary Ahlfors-regularity estimates for scales which are slightly smaller than $\operatorname{diam} E$. The Ahlfors-regularity estimates for scales all the way up to $\operatorname{diam} E$ then follow automatically from this.)

Let F and $\{\Sigma_j\}_{j\geq 0}$ be the Ahlfors-regular set and cubical patchwork provided by Proposition 7.6. Notice that

$$C^{-1} r_0 \leq \operatorname{diam}(S^* \cap B) \leq \operatorname{diam} F \leq 3 r_0,$$

because of Proposition 4.1 and (7.9). Let us also assume that $\operatorname{diam} Q \leq \frac{1}{100} r_0$ for all $Q \in \Sigma_j$, $j \geq 0$. This was true for the construction in the proof of Proposition 7.6, and it is easy to arrange in any case (by skipping the first few generations of cubes as needed). With these remarks we have that

(8.2) $\{\Sigma_j\}_{j\geq 0}$ satisfies (7.2), (7.3), and (7.10) with $R = r_0$
and with E replaced by S^*.

The substitution of S^* for E in (7.10) works because $F \subseteq \frac{3}{2}B$, while E and S^* agree in $\frac{5}{3}B$, which is strictly larger than $\frac{3}{2}B$. Let us also note that

(8.3) $$S^* \cap B \subseteq F \subseteq S^*,$$

because of (7.9) and the choice of E.

Let C_2 be a very large constant, the choice of which will be discussed later. For each $x \in F$ and each integer $j \geq 0$, define a set $T_j(x)$ by taking

$$T_j(x) = \bigcup \{Q \in \Sigma_j : Q \cap B(x, C_2 2^{-j} r_0) \neq \emptyset\}.$$

Note that $T_j(x)$ is a subset of F, since the Q's in Σ_j are contained in F by construction.

Let a d-plane P in $\mathbf{R}^n$ be given, and let π denote the orthogonal projection of $\mathbf{R}^n$ onto P. Also let C_1 and η be two other positive constants. They will be discussed later, but let us say now that C_1 will be fairly large, while η will be extremely small. Let $G_j = G_j(C_1, C_2, \eta)$ be the "good" set consisting of points $x \in F$ such that either

(8.4) $$\pi\big(T_j(x)\big) \supseteq P \cap B\big(\pi(x), C_1 2^{-j} r_0\big)$$

or

(8.5) there is a cubical set $R \in \Sigma_j$ such that $R \subseteq T_j(x)$

and R satisfies the following inequality:

(8.6) $$H^d\big(\pi(R)\big) \geq (1+2\eta) H^d(R) H^d\big(\pi(T_j(x))\big) H^d\big(T_j(x)\big)^{-1}.$$

Each of these alternatives represent favorable events in the stopping-time argument described in [**4**], i.e., it is useful to have them be true. We shall not be much more precise about their role in [**4**], except for the overall conclusions of [**4**].

MAIN LEMMA 8.7. *Notation and assumptions as above. For each choice of positive constants γ and C_1, there exist $C_2 > 0$ (large) and $\eta > 0$ (small) with the following property. Let $x \in F$ and $j \geq 0$ be given (and arbitrary). If*

(8.8) $$B\big(x, 2C_2 2^{-j} r_0\big) \subseteq \tfrac{1}{2} B$$

and

(8.9) $$H^d\big(\pi(T_j(x))\big) \geq \gamma H^d\big(T_j(x)\big),$$

then [we require that] $x \in G_j(C_1, C_2, \eta)$. The constants C_2 and η depend only on γ, C_1, n, k, and the constants for the cubical patchwork $\{\Sigma_j\}_j$ implicit in (8.2).

REMARK 8.10. The constants for $\{\Sigma_j\}_j$ do not cause any trouble here, because they are controlled by Proposition 7.6 (and, indirectly, by Proposition 4.1).

As we mentioned before, C_2 will be quite large, and in particular, every cube Q in Σ_j will have diameter much smaller than $C_2 2^{-j} r_0$. Thus (8.8) guarantees that $T_j(x)$ lies well inside $\tfrac{1}{2} B$. Let us also mention that the proof of Main Lemma 8.7 will not quite use all of the information we have about F and $\{\Sigma_j\}_j$, but only (8.3), the Ahlfors regularity of F, and the fact that $\{\Sigma_j\}_j$ is a cubical patchwork for F that satisfies (8.2).

The main lemma will be proved in Chapter 9, especially Section 9.2. For the moment we want to explain why it implies the local uniform rectifiability of S^*. We begin by recalling the theorem from [**4**] that we want to apply.

THEOREM 8.11. *Let F be a compact d-dimensional Ahlfors-regular set in $\mathbf{R}^n$, and let $\{\Sigma_j\}_{j \geq 0}$ be a cubical patchwork for F. Let $P \subseteq \mathbf{R}^n$ be a d-plane, and let π denote the orthogonal projection of $\mathbf{R}^n$ onto P. Also let $\gamma > 0$ be given. Then there is a constant $C_1 > 0$, depending only on the constants associated to $\{\Sigma_j\}_j$ (as in Definition 7.1), so that the following is true.*

Suppose that $j_0 \geq 0$ and $I_0 \in \Sigma_{j_0}$ are given, with

(8.12) $$H^d\big(\pi(I_0)\big) \geq 2\gamma H^d(I_0).$$

Also suppose that there are positive constants C_2 and η such that, with the same notations as before, $x \in G_j(C_1, C_2, \eta)$ whenever $j \geq j_0$ and $x \in I_0$ satisfy (8.9). Then there is a closed set $\Gamma \subseteq I_0$ such that

(8.13) $$H^d(\Gamma) \geq \theta H^d(I_0)$$

and

(8.14) $$|\pi(u) - \pi(v)| \geq M^{-1}|u - v| \quad \text{for all} \quad u, v \in \Gamma.$$

Here $\theta > 0$ and $M > 0$ depend only on n, γ, the constants for the cubical patchwork $\{\Sigma_j\}_j$, C_2, and η.

This is essentially nothing but a rephrasal of Theorem 1 on p. 77 of [4], modulo some minor adjustments. In [4] the result was stated for an unbounded regular set, while here we work with a compact regular set. This makes no real difference; none of the properties of the regular set F outside the given cubical set I_0 were ever used in the proof in [4], and the mapping z in [4] might as well not have been defined outside of I_0. In [4] the statements were normalized so that diam $I_0 \sim 1$, but this is also not significant. The "cubes" required in Condition Q on p. 76 of [4] are provided by the partitions Σ_j. Note that "dist$(x, \partial R)$" in (6) on p. 76 of [4] is a misprint, and should be replaced with "dist$(x, \Sigma \setminus R)$". This is compatible with the conditions (7.4) or (7.10) for our cubical sets. Notice also that the small number ϵ in Condition Q of [4] does not have any deep meaning: we can simply take $\epsilon = 2^{-N}$ for some large integer N, and the cubes in [4] will be obtained by keeping only the partitions $\Sigma_{j_0 + Nj}$, $j \geq 0$, rather than all of the Σ_k's. The main reason for taking ϵ small in [4] was to have the small boundary condition (6) on p. 76.

The number δ in [4] has been replaced by 2γ here, because we are using δ already in the quasiminimality condition from Chapter 1. The story about $G_j(C_1, C_2, \eta)$ and (8.9) corresponds to Condition (9) on p. 77 of [4]. Note that the requirement in [4] that Condition (9) apply to all x in the regular set — and not just the x's in I_0 — was unnecessary. In fact, the argument in [4] uses Condition (9) only when $x \in I_0$ also satisfies $T_j(x) \subseteq I_0$; otherwise the corresponding cubes were simply thrown away, and accounted for through the small boundary condition for I_0. (This is why we do not need to account for (8.8) in the statement of Theorem 8.11.)

Actually, we could have stated Theorem 8.11 with $I_0 = F$, and then the apparently more general version above would follow automatically. The present formulation was chosen for the convenience of the application to quasiminimizers.

Let us now explain how local uniform rectifiability of S^* can be derived from Theorem 8.11 and Main Lemma 8.7. Let U, S, and B be as in the beginning of this chapter. Choose F and the Σ_j's, $j \geq 0$, in the same manner as before. Let us assume for the moment that there is a constant $\gamma > 0$ with the property that, for each choice of cubical set $I_0 \in \Sigma_j$, $j \geq 0$, there is a d-plane P such that (8.12) holds.

In this situation we apply Theorem 8.11 to get a constant C_1 with a certain property. This constant C_1 depends on the constants associated to the cubical patchwork $\{\Sigma_j\}_j$, but the latter are under control, as in Remark 8.10. Using this choice of C_1, we then apply Main Lemma 8.7 to obtain particular values of C_2 and η.

Remember that x_0 is the center of our ball B, and that r_0 is the radius. Let j_0 be the smallest nonnegative integer with the property that if I_0 is the cube in Σ_{j_0}

that contains x_0, then
$$B(x, 2C_2 2^{-j_0} r_0) \subseteq \frac{1}{2} B$$
for every $x \in I_0$. Fix a d-plane P such that (8.12) holds with this choice of I_0, and let us use Theorem 8.11. The requirement that x lie in $G_j(C_1, C_2, \eta)$ for all $x \in I_0$ and all $j \geq j_0$ such that (8.9) holds is satisfied because of Main Lemma 8.7.

From Theorem 8.11 we obtain a closed set $\Gamma \subseteq I_0 \subseteq S^* \cap B$ that satisfies (8.13) and (8.14). Thus the restriction of π to Γ is bilipschitz, by (8.14), and
$$H^d(\Gamma) \geq \theta H^d(I_0) \geq C^{-1} r_0^d$$
for a constant C that we can control. (It might be really big, but it only depends on the correct parameters, and not on the initial choice of ball B, for instance.) This conclusion is exactly the kind of local uniform rectifiability property that we are after. It does not quite fit with the exact formulation of Theorem 2.11, and we shall deal with that in Chapter 10.

The argument just described relies on the assumption that we can find, for each cubical set I_0 in some Σ_j, a d-plane P so that (8.12) holds. To avoid this assumption we shall employ the trick discussed in Chapter 6. Using this we shall establish the following.

PROPOSITION 8.15. *Let S be a (U, k, δ)-quasiminimizer, and let $B = \overline{B}(x_0, r_0)$ be a closed ball centered on S^* such that $2B \subseteq U$ and $r_0 < \delta$. Then there is a compact subset Γ of $S^* \cap B(x_0, r_0)$ and a Lipschitz mapping $h : \Gamma \to \mathbf{R}^d$ such that*

(8.16) $$H^d(\Gamma) \geq C_3^{-1} r_0^d$$

and

(8.17) $$C_3^{-1} |u - v| \leq |h(u) - h(v)| \leq C_3 |u - v| \quad \text{for } u, v \in \Gamma,$$

where C_3 depends only on k and n.

The remainder of this chapter will be devoted to the proof of Proposition 8.15, given Main Lemma 8.7 and Theorem 8.11. Let S, B be as in the statement of Proposition 8.15, and let F, $\{\Sigma_j\}_{j \geq 0}$ be a regular set and a cubical patchwork for it, chosen in the same way as at the beginning of the chapter (i.e., through Proposition 7.6). We do not know that F has big projections yet, but we can find Lipschitz mappings with big images instead, as in the next lemma.

LEMMA 8.18. *For each $j_0 \geq 0$ and each cube $I_0 \in \Sigma_{j_0}$ such that $I_0 \subseteq B$, there is a C-Lipschitz mapping $h : \mathbf{R}^n \to \mathbf{R}^d$ such that*

(8.19) $$H^d\big(h(I_0)\big) \geq C^{-1} H^d(I_0),$$

where C depends only on n and k.

To prove this, let $I_0 \in \Sigma_{j_0}$ be given, with $I_0 \subseteq B$. Let us first check that there is a point $\xi \in I_0$ such that

(8.20) $$\operatorname{dist}(\xi, S^* \backslash I_0) \geq b \operatorname{diam} I_0,$$

where $b > 0$ depends only on k and n. This follows from (7.2) and (7.10) in (8.2). Specifically, (7.2) gives a lower bound for the mass of I_0, while (7.10) provides a (small) upper bound for the measure of the set of points in I_0 which lie close to $S^* \backslash I_0$. Thus there must be at least one point in I_0 which is not too close to $S^* \backslash I_0$, as in (8.20).

Let Q be the cube in $\mathbf{R}^n$ (in the standard sense) which is centered at ξ, has its sides parallel to the axes, and diameter equal to $b \operatorname{diam} I_0$. We may as well assume that $b \leq \frac{1}{2}$. This ensures that Q satisfies the hypothesis (5.2) in Proposition 5.1, because of our requirements on B from Proposition 8.15. From Proposition 5.1 we get a C-Lipschitz map $h : \mathbf{R}^n \to \mathbf{R}^d$ such that

$$(8.21) \qquad H^d\big(h(S^* \cap Q)\big) \geq C^{-1} (\operatorname{diam} I_0)^d.$$

This proves Lemma 8.18, because we know from (8.20) that $S^* \cap Q \subseteq I_0$.

Now we proceed to the proof of Proposition 8.15 itself. Let j_0 be a nonnegative integer, and let I_0 be an element of Σ_{j_0} with $I_0 \subseteq B$, where B is as in Proposition 8.15. We shall make specific choices later on, but for the moment let us proceed on this basis. Let h be as in Lemma 8.18, and set

$$(8.22) \qquad \widehat{S} = \{(x,y) \in \mathbf{R}^n \times \mathbf{R}^d : x \in S \text{ and } y = h(x)\}.$$

Also put

$$\widehat{S}^* = \{(x,y) \in \mathbf{R}^n \times \mathbf{R}^d : x \in S^* \text{ and } y = h(x)\}.$$

Thus $\widehat{S}^*$ is the same as $\big(\widehat{S}\big)^*$, where the latter is defined through (1.12) (with $\widehat{U} = U \times \mathbf{R}^d$), i.e., as the intersection of $\widehat{U}$ with the support of H^d restricted to $\widehat{S}$. This uses the fact that $h : \mathbf{R}^n \to \mathbf{R}^d$ is Lipschitz.

Denote by $\widehat{F}$ the graph of h over F, i.e.,

$$(8.23) \qquad \widehat{F} = \{(x,y) \in \mathbf{R}^n \times \mathbf{R}^d : x \in F \text{ and } y = h(x)\}.$$

For each $j \geq 0$ and $Q \in \Sigma_j$, put

$$(8.24) \qquad \widehat{Q} = \{(x,y) \in \mathbf{R}^n \times \mathbf{R}^d : x \in Q \text{ and } y = h(x)\},$$

and set

$$(8.25) \qquad \widehat{\Sigma}_j = \{\widehat{Q} : Q \in \Sigma_j\}.$$

With these choices $\widehat{F}$ is an Ahlfors-regular set of dimension d, and $\{\widehat{\Sigma}_j\}_{j \geq 0}$ is a cubical patchwork for $\widehat{F}$. Moreover,

$$(8.26) \qquad \widehat{B} \cap \widehat{S}^* \subseteq \widehat{F} \subseteq \widehat{S}^*$$

(as in (8.3)), where $\widehat{B}$ is the ball in $\mathbf{R}^n \times \mathbf{R}^d$ with center $\big(x_0, h(x_0)\big)$ and radius r_0, and $\{\widehat{\Sigma}_j\}_{j \geq 0}$ satisfies (7.10) with $E = \widehat{S}^*$. (Compare with (8.2).) These assertions follow from the analogous statements for F, $\{\Sigma_j\}_j$, and from the fact that $h : \mathbf{R}^n \to \mathbf{R}^d$ is C-Lipschitz. In each case one has estimates that depend only on k and n.

Let P denote the d-plane $\{0\} \times \mathbf{R}^d$ in $\mathbf{R}^n \times \mathbf{R}^d$, and let π be the obvious projection from $\mathbf{R}^n \times \mathbf{R}^d$ onto P. Our choice of h ensures that

$$(8.27) \qquad H^d\big(\pi(\widehat{I}_0)\big) = H^d\big(h(I_0)\big) \geq 2\gamma H^d\big(\widehat{I}_0\big),$$

where $\gamma > 0$ depends only on k and n.

We now apply Theorem 8.11 to the set $\widehat{F}$ and cubical patchwork $\{\widehat{\Sigma}_j\}_{j \geq 0}$, with $\gamma > 0$ chosen as above. From this we obtain a new constant C_1, which depends only on k and n, and which has the property described in the statement of Theorem 8.11.

By Proposition 6.1, $\widehat{S}$ is a $(\widehat{U}, \widehat{k}, \delta)$-quasiminimizer in $\mathbf{R}^n \times \mathbf{R}^d$, where $\widehat{U} = U \times \mathbf{R}^d$ and $\widehat{k}$ depends only on k and n. Thus we are in the same situation as described at the beginning of this chapter, except that S, S^*, F, and $\{\Sigma_j\}_j$ have been replaced with $\widehat{S}$, $\widehat{S}^*$, $\widehat{F}$, and $\{\widehat{\Sigma}_j\}_j$. This permits us to apply Lemma 8.7,

8. A STOPPING-TIME ARGUMENT

with $\gamma > 0$ as in (8.27) and with the constant C_1 that we just obtained from Theorem 8.11. This leads to new constants $C_2 > 0$ and $\eta > 0$, which still depend only on k and n.

So far everything that we have said applies to any choice of $j_0 \geq 0$ and $I_0 \in \Sigma_{j_0}$ with $I_0 \subseteq B$. In particular, the constants γ, C_1, C_2, and η do not depend on j_0 and I_0. The Lipschitz mapping h from Lemma 8.18 does depend on j_0 and I_0, and the sets $\widehat{S}$, $\widehat{F}$, etc., do as well, but this does not cause any trouble, because we have uniform bounds.

We are now ready to make specific choices of j_0 and I_0 (in order to prove Proposition 8.15). Once we select j_0, we simply take I_0 to be the cubical set in Σ_{j_0} that contains the center x_0 of the given ball B. (See Proposition 8.15.) For j_0 we take a nonnegative integer which is large enough so that

$$(8.28) \qquad B(\widehat{x}, 2C_2 2^{-j_0} r_0) \subseteq \tfrac{1}{2}\widehat{B} \quad \text{for all } \widehat{x} \in \widehat{I}_0.$$

We can do this in such a way that j_0 is bounded by a constant that depends only on C_2, the constants for the cubical patchwork $\{\Sigma_j\}_j$, and on a bound for the Lipschitz constant for the mapping h in Lemma 8.18. There is no vicious circle here, in this bound for j_0, because we made certain before that the constant C_2 and the bound for the Lipschitz constant of h could be chosen independently of j_0 and I_0.

Thus we can safely take a sufficiently large (but controlled) value of j_0, and then define I_0, h, $\widehat{F}$, etc., as above. We now apply Theorem 8.11, but with $\widehat{F}$, $\{\widehat{\Sigma}_j\}_{j\geq 0}$, and $\widehat{I}_0$. Notice that the analogue of (8.12) for $\widehat{I}_0$ is the same as (8.27). Also, Main Lemma 8.7 (applied to $\widehat{F}$ and $\{\widehat{\Sigma}_j\}_{j\geq 0}$) guarantees that $\widehat{x}$ lies in $G_j(C_1, C_2, \eta)$ whenever $j \geq j_0$ and $\widehat{x} \in \widehat{I}_0$ satisfy the analogue of (8.9). This also uses (8.28), which gives the analogue of (8.8) in this situation.

In other words, the "hypotheses" of the conclusions of Theorem 8.11 hold in this situation, and so we obtain the existence of a closed subset $\widehat{\Gamma}$ of $\widehat{I}_0$ such that

$$(8.29) \qquad H^d(\widehat{\Gamma}) \geq \theta H^d(I_0),$$

and on which π is M-bilipschitz. The constants θ and M can be taken to depend only on k and n, since that is true of the various other constants involved here. Let Γ be the projection of $\widehat{\Gamma}$ on $\mathbf{R}^n$ (using the obvious projection from $\mathbf{R}^n \times \mathbf{R}^d$ onto $\mathbf{R}^n$). It is clear from the construction that $\Gamma \subseteq I_0 \subseteq F \cap B(x_0, r_0) = S^* \cap B(x_0, r_0)$, i.e., because $\widehat{\Gamma} \subseteq \widehat{I}_0$ and so forth. Since $\widehat{\Gamma}$ is the graph of the Lipschitz function h over Γ, we have that

$$H^d(\Gamma) \geq A^{-1} H^d(\widehat{\Gamma})$$

for a suitable constant A^{-1} (that depends only on d and the Lipschitz constant for h). On the other hand, $H^d(I_0) \geq C^{-1}(2^{-j_0} r_0)^d$, since $I_0 \in \Sigma_{j_0}$. More precisely, this uses (7.2) and the fact that the diameter of F is comparable in size to r_0. The latter comes from the way that we chose F, which follows the construction at the beginning of the chapter, as indicated just after the statement of Proposition 8.15. We also know that j_0 is uniformly bounded, as discussed just after (8.28). Combining these inequalities we conclude that (8.16) holds with this choice of Γ, if the constant C_3 in (8.16) is large enough.

This leaves (8.17). We already know that h is Lipschitz with a bounded constant, as in Lemma 8.18, and so we only have to worry about the lower bound in

(8.17). For this one uses the (trivial) formula

(8.30) $$h(u) = \pi\bigl(u, h(u)\bigr) \qquad \text{for } u \in \Gamma,$$

together with the fact that π is bilipschitz on $\widehat{\Gamma}$ (= the graph of h over Γ).

This completes the proof of Proposition 8.15, modulo Main Lemma 8.7. We shall discuss the local uniform rectifiability properties of S^* further in Chapter 10 (i.e., in other formulations), after proving Main Lemma 8.7 in Chapter 9.

CHAPTER 9

Proof of Main Lemma 8.7

The main engine of this proof will be a variant of a deformation lemma in [4], which we shall state and prove separately.

9.1. A general deformation result

We begin with some notation. Let P be a d-plane in $\mathbf{R}^n$ (with $0 < d < n$), and let V be an $(n-d)$-plane orthogonal to P. We shall think of $\mathbf{R}^n$ as being identified with the Cartesian product of P and V, and generic elements z of $\mathbf{R}^n$ will often be written as $z = (x, y)$, with $x \in P$ and $y \in V$. Let π and h denote the orthogonal projections from $\mathbf{R}^n$ onto P and V, respectively (so that $\pi(z) = x$ and $h(z) = y$ when $z = (x, y)$).

We also ask that P come equipped with a fixed system of orthonormal co-ordinates. Whenever we refer to a "cube" in P, we shall mean a closed cube with sides parallel to the axes determined by these co-ordinates. By the "unit cube" we mean the one that corresponds to $[0, 1]^d \subseteq \mathbf{R}^d$ in these co-ordinates. Let $\{A_i\}_{i \in I}$ denote the tiling of P by integer translates of this unit cube. Thus P is covered by the union of the A_i's, while the "interiors" of the A_i's (in P) are pairwise disjoint.

Let P_0 denote the subset of P which corresponds to the subset $[-N, N]^d$ of $\mathbf{R}^d$ with respect to the given co-ordinates. Here N is a large integer (to be specified later). We shall also assume that we are given a subset V_0 of V, with V_0 either a cube or a ball. In practice, the size of V_0 will be comparable to N, but for the purposes of Proposition 9.6 below it is enough for V_0 to be convex, say. Set

$$(9.1) \qquad Q_0 = P_0 \times V_0 \subseteq \mathbf{R}^n.$$

Let E_0 be a compact subset of Q_0, and set $I_0 = \{i \in I : A_i \subseteq P_0\}$ (so that the cubes A_i, $i \in I_0$, cover P_0). Put

$$(9.2) \qquad I_1 = \{i \in I_0 : \text{there is an } x_i \in \text{int}(A_i) \text{ such that}$$
$$E_0 \cap \pi^{-1}(x_i) = \emptyset\}$$

and $I_2 = I_0 \backslash I_1$. We shall assume that

$$(9.3) \qquad \text{for each } i \in I_0, \text{ there is a finite set } \Xi(i) \subseteq E_0 \cap \pi^{-1}(A_i)$$
$$\text{with at most } C_0 \text{ elements such that } \text{dist}(z, \Xi(i)) \leq 1$$
$$\text{for every } z \in E_0 \cap \pi^{-1}(A_i),$$

that

$$(9.4) \qquad I_1 \text{ is not empty,}$$

and that

(9.5) for each $i \in I_2$, there is a point $(x_i, y_i) \in E_0$ such that
$x_i \in \mathrm{int}(A_i)$ and $|y - y_i| \leq 1$ for all $y \in V$ which satisfy
$(x_i, y) \in E_0$.

(Note that $\Xi(i)$ may be empty, for any given i.)

These assumptions say something about the extent to which E_0 behaves approximately like a graph, or a union of pieces of graphs, over P_0. More precisely, condition (9.3) says that there are at most a bounded number of "clumps" in E_0 which lie above any given A_i. One does not know or assume anything about the distances between these clumps a priori. Inside each A_i there is at least one point x_i which has only one "clump" in E_0 lying directly above it, by (9.5). There is at least one point in one A_i which has nothing above it, because of (9.4).

The next proposition provides a way to make a controlled deformation of E_0 onto a set of lower dimension. This deformation will not be a local one — points may be moved far away from where they started — but it will never move points in E_0 too far from E_0.

Conditions (9.4) and (9.5) are crucial for this construction. The "hole" given by (9.4) provides a place to start. One would then like to systematically push away from that hole. This would not work if E_0 looked like some kind of parabola, bending away from the hole in two different directions, but that possibility is prevented by (9.5). It would also not work if E_0 looked like some kind of flat sheet which was completely parallel to the hole corresponding to (9.4), and this avoided by (9.3).

In the end, one does not eliminate E_0 entirely, but one reduces it to a lower-dimensional residue that lies above the boundaries of the cubes A_i, $i \in I$. This permits one to conclude that E_0 was trivial for certain purposes of d-dimensional topology. In practice one often turns this around, and says that one of (9.3), (9.4), or (9.5) must be false under suitable conditions of topological nondegeneracy.

Proposition 9.6 is similar to some fairly standard constructions in topology, but part of the point here is that we need to be a bit careful about uniform bounds.

PROPOSITION 9.6. *Let P_0, V_0, and E_0 be as above, so that (in particular) (9.3), (9.4), and (9.5) hold for some constant $C_0 > 0$. Then there is a family of mappings Φ_t, $0 \leq t \leq 1$, on $\mathbf{R}^n$, with the following properties:*

(9.7) $(t, z) \longrightarrow \Phi_t(z)$ *is Lipschitz as a mapping from*
$[0, 1] \times \mathbf{R}^n$ *to* $\mathbf{R}^n$;

(9.8) $\Phi_0(z) = z$ *for all* $z \in \mathbf{R}^n$;

(9.9) $\Phi_t(z) = z$ *for all t when* $\mathrm{dist}(z, Q_0) \geq d + 3$;

(9.10) *if $\Phi_t(z) \neq z$ for some t, then* $\mathrm{dist}(z, E_0) \leq C$, *and*
$\mathrm{dist}(\Phi_u(z), E_0) \leq C$ *for all* $u \in [0, 1]$;

(9.11) $\Phi_t(E_0) \subseteq Q_0$ *for all t;*

(9.12) $\Phi_1(E_0)$ *has finite $(d - 1)$-dimensional Hausdorff measure*
(and hence H^d-measure 0);

(9.13) Φ_t *is C-Lipschitz on* $\mathbf{R}^n \backslash (P_0 \times V)$.

Here C is a constant that depends only on C_0 and n.

Note that we do not assert here the existence of a uniform bound for the Lipschitz norm of Φ_t on Q_0 (in (9.7)), or even on E_0. The construction behind the proof does not give such a bound either.

When we apply Proposition 9.6 to prove Main Lemma 8.7 (in Section 9.2), we shall take E_0 to be $Q_0 \cap S^*$ for a quasiminimizing set S, and (9.12) will tell us that Φ_1 eliminates E_0 in terms of d-dimensional measure. The additional regularity provided by (9.13) will be useful for showing that Φ_1 does not increase the measure of S^* near the boundary of Q_0 too much. Some of the properties listed in Proposition 9.6 are stronger than needed for this particular application, but they are easy to get and useful in other contexts.

The proof of Proposition 9.6 will be very similar to the argument given for Lemma 16 on p. 108 of [**4**], but the treatment here will be more detailed.

The construction of $\{\Phi_t\}$ will be accomplished in three main steps. In the first step we define a mapping ϕ_1, which we shall later take to be $\Phi_{\frac{1}{3}}$. This mapping ϕ_1 will move points only in directions parallel to P. For each of the cubes A_i in P, ϕ_1 will have the general tendency of moving points in $\pi^{-1}(A_i)$ to $\pi^{-1}(\partial A_i)$, at least to the extent that it is possible to do so.

The first step of the construction. To define ϕ_1 we proceed as follows. For each $i \in I_1$, let x_i be as in (9.2), and, for each $i \in I_2$, let x_i and y_i be as in (9.5). Denote by θ_i the "radial" projection from $A_i \backslash \{x_i\}$ onto ∂A_i (centered at x_i). Thus θ_i is Lipschitz on any compact subset of $A_i \backslash \{x_i\}$, and it fixes every element of ∂A_i. We want to choose $\phi_1 : \mathbf{R}^n \to \mathbf{R}^n$ so that it satisfies:

(9.14) $\quad \phi_1$ is Lipschitz;

(9.15) $\quad h \circ \phi_1 = h$ on all of $\mathbf{R}^n$;

(9.16) $\quad \phi_1(z) = z$ when $\pi(z) \in P \backslash P_0$, and also when $\operatorname{dist}(z, E_0) \geq 1$;

(9.17) $\quad \pi(\phi_1(x,y)) = \theta_i(x)$ whenever $(x,y) \in E_0$ satisfies

(9.18a) $\quad\quad x \in \partial A_i$ for some $i \in I_0$, or

(9.18b) $\quad\quad x \in A_i$ for some $i \in I_1$, or

(9.18c) $\quad\quad x \in A_i$ and $|y - y_i| \geq 2$ for some $i \in I_2$;

and finally

(9.19) $\quad\quad\quad \pi(\phi_1(x,y)) \in A_i$ when $x \in A_i$.

We first put

$$\phi_1(z) = z \quad \text{whenever } \pi(z) \in P \backslash P_0$$
$$\text{or } \pi(z) \in \partial A_i \text{ for some } i \in I_0.$$

This is compatible with (9.17), because $\theta_i(x) = x$ for all $x \in \partial A_i$, by construction.

Given $i \in I_0$, let W_i denote the set of $z \in E_0$, $z = (x,y)$, such that z satisfies (9.18b) or (9.18c) (according to whether i lies in I_1 or I_2). The conditions (9.15) and (9.17) determine the values of ϕ_1 uniquely on each W_i, and so we can think of ϕ_1 as now being extended to all of the W_i's as well. Let us check that the restriction of ϕ_1 to $\pi^{-1}(\partial A_i) \cup W_i$ is Lipschitz for each $i \in I_0$. We only have to worry about $\pi \circ \phi_1$, since (the remaining piece) $h \circ \phi_1$ is equal to h, as in (9.15). Notice that $\pi(W_i)$ is a compact subset of A_i that necessarily omits x_i, because of (9.2) and (9.5) (for i in I_1 and I_2, respectively). This implies that θ_i is Lipschitz on $\partial A_i \cup \pi(W_i)$, and hence that $\pi \circ \phi_1$ is Lipschitz on $\pi^{-1}(\partial A_i) \cup W_i$, because of (9.17).

Set $U_i = E_0 \cap \pi^{-1}(A_i)$. Thus $U_i = W_i$ when $i \in I_1$, but this can fail when $i \in I_2$. For each $i \in I_2$, let us extend ϕ_1 as a mapping defined so far on $\pi^{-1}(\partial A_i) \cup W_i$ to a mapping defined on the (potentially larger) set $\pi^{-1}(\partial A_i) \cup U_i$. It is easy to do this in such a manner that the extension remains Lipschitz, the condition $h \circ \phi_1 = h$ is preserved, and $\pi(\phi_1(z)) \in A_i$ when $z \in U_i$, since these features hold already on $\pi^{-1}(\partial A_i) \cup W_i$.

Thus we have ϕ_1 defined as a Lipschitz mapping on

$$\pi^{-1}(\partial A_i) \cup \left(E_0 \cap \pi^{-1}(A_i)\right)$$

for each $i \in I_0$, and with $\phi_1(z) = z$ whenever $\pi(z) \in \partial A_i$ for any i. This compatibility condition ensures that ϕ_1 is actually Lipschitz on the union of these individual sets. In fact, we have now defined ϕ_1 for all z such that

$$\pi(z) \in P \backslash P_0, \quad \text{or } \pi(z) \in \partial A_i \text{ for some } i, \quad \text{or } z \in E_0,$$

and ϕ_1 is Lipschitz as a mapping on this whole set.

In accordance with (9.16), we now set $\phi_1(z) = z$ also for $z \in \mathbf{R}^n$ such that $\text{dist}(z, E_0) \geq 1$. It is not hard to see that this new extension of ϕ_1 is also Lipschitz. Finally, we define ϕ_1 on all of $\mathbf{R}^n$ by taking another Lipschitz extension. We demand that this extension also be compatible with (9.15) and (9.19), but this is easy to arrange, because we had the same kind of compatibility at every previous step. More precisely, to get an extension for ϕ_1, it is enough to choose extensions for $h \circ \phi_1$ and $\pi \circ \phi_1$ separately, and then combine them afterwards. In this manner, compatibility with (9.15) and (9.19) "decouples", and they become independent properties for the extensions of $h \circ \phi_1$ and $\pi \circ \phi_1$ on their own (without involving the other). For $h \circ \phi_1$, one can simply take the extension to $\mathbf{R}^n$ to be equal to h everywhere, as in (9.15). For $\pi \circ \phi_1$, it is not hard to see that there is a Lipschitz extension which is compatible with (9.19). One can choose the extension on each set $\pi^{-1}(A_i)$, $i \in I_0$, separately, for instance. We have already defined $\pi \circ \phi_1$ on the complement of the union of these sets, and on the boundaries of these sets, and this ensures the consistency of the individual extensions to the sets $\pi^{-1}(A_i)$, $i \in I_0$. That is, one has consistency of these individual extensions with each other, and with the rest of $\pi \circ \phi_1$ on $\pi^{-1}(P \backslash P_0)$.

This completes the construction of the mapping $\phi_1 : \mathbf{R}^n \to \mathbf{R}^n$ that satisfies (9.14)–(9.19). Notice that we already have no control on the Lipschitz constant for ϕ_1 on E_0. The Lipschitz behavior that we do have depends on (9.17) and the fact that $\pi(W_i)$ is a compact subset of A_i which omits x_i, but we have no bounds (from below) for the distance between $\pi(W_i)$ and x_i, while the Lipschitz norm of θ_i blows up as one approaches x_i.

The second step of the construction. Set

(9.20) $$E_1 = \phi_1(E_0) \subset Q_0,$$

and

(9.21) $$\Phi_t(z) = (1 - 3t)z + 3t\phi_1(z) \quad \text{for } 0 \leq t \leq \tfrac{1}{3} \text{ and } z \in \mathbf{R}^n.$$

Our next goal is to define a mapping $\phi_2 : \mathbf{R}^n \to \mathbf{R}^n$ that contracts the set E_1 about as much as it can in the y-direction. Specifically, $E_2 = \phi_2(E_1)$ will have the good feature that $\pi^{-1}(x) \cap E_2$ has at most C points for each $x \in P$, where C depends

9.1. A GENERAL DEFORMATION RESULT

only on n and our original constant C_0. This mapping ϕ_2 will only move points in directions parallel to V, so that

$$\pi \circ \phi_2 = \pi \quad \text{on } \mathbf{R}^n. \tag{9.22}$$

The definition of ϕ_2 will be somewhat more intricate than that of ϕ_1.

Put

$$\Xi = \bigcup_{i \in I_0} \Xi(i), \tag{9.23}$$

where $\Xi(i)$ is as in (9.3). Without loss of generality, we may assume that the point (x_i, y_i) in (9.5) lies in $\Xi(i)$ for each $i \in I_2$ (i.e., otherwise we simply add (x_i, y_i) to $\Xi(i)$).

For each $\xi \in \Xi$, write $\xi = (x_\xi, y_\xi)$, and set

$$H(\xi) = \{(x,y) \in \mathbf{R}^n : |x - x_\xi| + |y - y_\xi| \leq d+2\}. \tag{9.24}$$

Also put $H = \bigcup_{\xi \in \Xi} H(\xi)$ and

$$H^x = \{y \in V : (x,y) \in H\} \tag{9.25}$$

for each $x \in P$. We shall use certain functions $d_\xi(x, \cdot)$ to measure (distorted) distances to the points y_ξ in V, $\xi \in \Xi$. Put

$$d_\xi(x,y) = \inf\{H^1(\Gamma \backslash H^x) : \Gamma \text{ is a rectifiable arc in } V \tag{9.26}$$
$$\text{that connects } y \text{ to } y_\xi\}.$$

Here $x \in P$, $y \in V$, and $\xi \in \Xi$ are arbitrary, and $H^1(\cdot)$ denotes 1-dimensional Hausdorff measure (as defined in (0.1), (0.2)). (Let us emphasize that $d_\xi(x,y)$ is a measurement of distance from y to y_ξ which depends on x as a kind of parameter, rather than a measurement of distance between x and y, as the notation might suggest.)

Let us check that

$$d_\xi(x,y) = d_\xi(x, y_{\xi'}) \quad \text{when } \xi, \xi' \in \Xi \text{ and } (x,y) \in H(\xi'). \tag{9.27}$$

If $(x,y) \in H(\xi')$, then both y and $y_{\xi'}$ lie in H^x, and in fact the whole line segment connecting y to $y_{\xi'}$ lies in H^x. The function $d_\xi(x, \cdot)$ is necessarily constant on any line segment in H^x, and (9.27) follows.

It is easy to see that $d_\xi(x,y)$ is 1-Lipschitz as a function of y. This follows from the definition (9.26), and it does not depend on the particular choice of H^x. Let us use the special structure of the H^x's to show that $d_\xi(x,y)$ is C-Lipschitz as a function of x (for each fixed y), where C depends only on C_0 and n. (Remember that C_0 comes from (9.3).)

We should begin by taking a closer look at the sets H^x. For each $x \in P$, one can realize H^x as the union of at most $C(n) \cdot C_0$ closed balls of radius $\leq d+2$. This uses the fact that $\Xi(i)$ contains at most C_0 elements for each i, and the observation that $H(\xi)$ does not contribute to H^x when $|x - x_\xi| > d+2$. In particular, $H(\xi)$ does not contribute to H^x when $\xi \in A_i$ and $\mathrm{dist}(x, A_i) > d+2$ (so that only boundedly many i's or ξ's are really involved for any fixed x).

As one moves x, the radii of the balls of which H^x is composed change in a 1-Lipschitz manner (by the definition (9.24)). The centers remain the same (i.e., the y_ξ's), except when the constituent balls suddenly appear or disappear as some $H(\xi)$ moves in or out of view.

Fix $\xi \in \Xi$, $y \in V$, and a pair of points x, $x' \in P$, and let us try to prove an inequality like

$$d_\xi(x',y) \leq d_\xi(x,y) + C|x-x'|.$$

Let Γ be a rectifiable arc in V that connects y to y_ξ and which approximately realizes the infimum in (9.26) (for x). We want to show that Γ can be modified to give an arc Γ' that also connects y to y_ξ and which satisfies

$$H^1(\Gamma'\backslash H^{x'}) \leq H^1(\Gamma\backslash H^x) + C|x-x'|.$$

In the transition from x to x', some of the balls of which H^x is composed may have become larger. This is perfectly fine for the desired inequality, because this leads to a decrease in $H^1(\Gamma\backslash H^x)$. However, some of the balls might get smaller, so that portions of Γ that were covered by H^x become exposed for $H^{x'}$. Imagine, for instance, that B is one of the balls in H^x that is replaced by a slightly smaller ball B' in $H^{x'}$ (and with the same center, by construction). To control the possible increase in $H^1(\Gamma\backslash H^x)$, consider the maximal subarc α of Γ with *endpoints* contained in B. There is a unique single such arc, and it contains all the separate pieces which lie in B. (One might first think in terms of working with the maximal subarcs of Γ which are contained in B, but this is simpler and works better.) Imagine that we we cut α out from Γ, and replace it with an arc α' that begins at the same point that α does, goes straight from there to the center of B (i.e., along a radius of B), and then goes straight from the center of B to the endpoint of α. If we do this, then $H^1(\alpha\backslash H^{x'})$ will be at most twice the difference in the radii of B and B'. In particular, $H^1(\alpha\backslash H^{x'})$ will be bounded by $2|x-x'|$, because of our earlier discussion about the 1-Lipschitz nature (as a function of x) of the radii of the balls that make up H^x. (Note that it is sometimes much better in terms of length to use this replacement for α – heading directly towards the center of B and back out again – rather than simply replacing α by a straight line segment between its endpoints. This is because such line segments could be approximately tangent to B, so that the amount of length exposed in the passage from B to B' is much more than a constant times $|x-x'|$.)

We can do this for each of the balls B of which H^x is composed, including the ones that disappear altogether. Strictly speaking, one can think of performing this operation sequentially one ball after another, since different balls might overlap. (Operating sequentially is fine for the estimates that we want in the end too, as one can check.) In this manner we obtain a modification Γ' of Γ with the same endpoints as before, and with $H^1(\Gamma'\backslash H^{x'})$ increased over $H^1(\Gamma\backslash H^x)$ by an amount that is bounded by the number of balls in H^x times a constant multiple of $|x-x'|$. This is exactly what we want, since we know that the number of balls that make up H^x is bounded by a multiple of C_0.

Thus we can bound $d_\xi(x',y)$ by the sum of $d_\xi(x,y)$ and $C|x-x'|$, where C depends only on C_0 and n. This proves that $d_\xi(x,y)$ is C-Lipschitz as a function of x, for fixed y and ξ (since we can just as well do the same with the roles of x, x' reversed). Combining this with our earlier observation that $d_\xi(x,y)$ is 1-Lipschitz in y, we conclude that

(9.28) $\qquad d_\xi$ is C-Lipschitz as a function of $(x,y) \in \mathbf{R}^n$.

9.1. A GENERAL DEFORMATION RESULT

Let us now record a couple of simple lower bounds for $d_\xi(x,y)$. The first says that there is a constant C' such that

(9.29) $$d_\xi(x,y) \geq 1 \quad \text{when} \quad |y - y_\xi| \geq C',$$

where C' depends only on C_0 and n. This is not hard to verify, using the fact that H^x can be realized as a union of (only) a bounded number of balls of radius $\leq d+2$. For instance, this permits one to choose C' large enough so that for each y_ξ there is a radius $R \leq C' - 1$ such that

$$H^x \cap \{z \in V : R \leq |z - y_\xi| \leq R+1\} = \emptyset,$$

and this is sufficient to give (9.29).

The next lower bound says that

(9.30) $$d_\xi(x,y) \geq 1 \quad \text{when} \quad \text{dist}((x,y), E_0) \geq d+3.$$

To prove this, it suffices to show that $\text{dist}(u, E_0) \leq d+2$ whenever $u \in H$, because of the definition (9.26) of $d_\xi(x,y)$. Let $u \in H$ be arbitrary, and choose $\xi' \in \Xi$ so that $u \in H(\xi')$. Then $\xi' \in E_0$, by construction, and $|u - \xi'| \leq d+2$, as in (9.24). Thus $\text{dist}(u, E_0) \leq d+2$, and (9.30) follows.

In order to define the mapping ϕ_2 we shall need a cut-off function and some other auxiliary functions which will serve as (variable) coefficients. Set

(9.31) $$\chi_\xi(x) = \begin{cases} 0 & \text{when } |x - x_\xi| \geq d+3 \\ 1 & \text{when } |x - x_\xi| \leq d+2 \\ d+3 - |x - x_\xi| & \text{otherwise} \end{cases}$$

and

(9.32) $$\lambda_\xi(x,y) = \chi_\xi(x) \, \text{Max}(0, 1 - d_\xi(x,y))$$

for all $\xi \in \Xi$ and $(x,y) \in P \times V$, and put

(9.33) $$\lambda_0(x,y) = 1 - \text{Max}_{\xi \in \Xi} \lambda_\xi(x,y).$$

Note that $\lambda_0(x,y)$ is always nonnegative (as is $\lambda_\xi(x,y)$, for that matter). Also set

(9.34) $$\sigma(x,y) = \lambda_0(x,y) + \sum_{\xi \in \Xi} \lambda_\xi(x,y).$$

Thus $\sigma(x,y) \geq 1$ by construction. We define "normalized coefficient functions" by

(9.35) $$\lambda'_\xi(x,y) = \sigma(x,y)^{-1} \lambda_\xi(x,y)$$

and

(9.36) $$\lambda'_0(x,y) = \sigma(x,y)^{-1} \lambda_0(x,y).$$

For each point (x,y) in $\mathbf{R}^n$, we have that $\lambda_\xi(x,y) = 0$ for all but C choices of $\xi \in \Xi$, where C depends only on C_0 and n. This follows from the definition (9.31) of $\chi_\xi(x)$ for localization in x (which ensures that only boundedly many $\Xi(i)$'s are involved), and then (9.3) to bound the number of elements in each $\Xi(i)$ separately.

Let us apply this to check that $\lambda'_\xi(x,y)$ and $\lambda'_0(x,y)$ are Lipschitz in (x,y) (on all of $\mathbf{R}^n$), and with norm bounded by a constant that depends only on C_0 and n. It is clear from (9.31) that $\chi_\xi(x)$ is 1-Lipschitz in x, and hence that $\lambda_\xi(x,y)$ is Lipschitz in x and y, and with bounded constant, because of (9.28). This also uses the boundedness of the quantities in the definition (9.32) of $\lambda_\xi(x,y)$. From here we get that $\lambda_0(x,y)$ is Lipschitz with bounded constant, and we can derive the same conclusion for $\sigma(x,y)$, since we know that no more than a bounded number of terms in the sum in (9.34) can be nonzero for any fixed (x,y) (as in the paragraph above). From here the desired Lipschitz conditions for $\lambda'_\xi(x,y)$ and $\lambda'_0(x,y)$ follow from straightforward computation (i.e., of differences of products), since $\sigma(x,y) \geq 1$ and $\lambda_\xi(x,y)$, $\lambda_0(x,y)$ take values in $[0,1]$.

We now define ϕ_2 on $\mathbf{R}^n$ by (9.22) and

$$(9.37) \qquad h\big(\phi_2(x,y)\big) = \lambda'_0(x,y)\, y + \sum_{\xi \in \Xi} \lambda'_\xi(x,y)\, y_\xi.$$

Let us first compute the values of ϕ_2 in some special situations.

Let us check that

$$(9.38) \qquad \phi_2(x,y) = \phi_2(x, y_{\xi'}) \quad \text{when} \quad (x,y) \in H(\xi').$$

It suffices to show that $h\big(\phi_2(x,y)\big) = h\big(\phi_2(x, y_{\xi'})\big)$, since $\pi\big(\phi_2(x,v)\big) = x$ for all $v \in V$, by (9.22). Since $(x,y) \in H(\xi')$, we have that

$$d_\xi(x,y) = d_\xi(x, y_{\xi'}) \quad \text{for all} \ \xi \in \Xi,$$

as in (9.27). (Note the difference in roles of ξ and ξ' here.) In particular,

$$d_{\xi'}(x,y) = d_{\xi'}(x, y_{\xi'}) = 0$$

(i.e., for the case where $\xi' = \xi$). We also have that $\chi_{\xi'}(x) = 1$, because of (9.31), the assumption that $(x,y) \in H(\xi')$, and the definition (9.24) of $H(\xi')$. This implies that

$$\lambda_{\xi'}(x,y) = \lambda_{\xi'}(x, y_{\xi'}) = 1,$$

and hence that $\lambda_0(x,y) = \lambda_0(x, y_{\xi'}) = 0$. To prove (9.38) we are now reduced to showing that

$$\lambda'_\xi(x,y) = \lambda'_\xi(x, y_{\xi'}) \quad \text{for all} \ \xi \in \Xi,$$

because of the definition (9.37) of $h \circ \phi_2$. To get this it is enough to verify that

$$\lambda_\xi(x,y) = \lambda_\xi(x, y_{\xi'}) \quad \text{for all} \ \xi \in \Xi,$$

(i.e., λ instead of λ'), because of the vanishing of λ_0. Since $\chi_\xi(x)$ doesn't depend on y, the definition (9.32) of λ_ξ permits us to derive this last family of equalities from the observation above that $d_\xi(x,y) = d_\xi(x, y_{\xi'})$ for all $\xi \in \Xi$.

A similar argument implies that

$$(9.39) \qquad \phi_2(z) = z \quad \text{when} \quad \text{dist}(z, E_0) \geq d+3.$$

Specifically, if $z = (x,y)$ satisfies $\text{dist}(z, E_0) \geq d+3$, then $d_\xi(x,y) \geq 1$ for all $\xi \in \Xi$, by (9.30), and hence $\lambda_\xi(x,y) = 0$ for all ξ, by (9.32). This yields $\lambda_0(x,y) = 1$, $\sigma(x,y) = 1$, $\lambda'_\xi(x,y) = 0$ for all ξ, and $\lambda'_0(x,y) = 1$. Putting this into (9.37) gives $h\big(\phi_2(x,y)\big) = y$, from which (9.39) follows (because of (9.22) for $\pi\big(\phi_2(x,y)\big) = x$).

Next, let us check that

$$(9.40) \qquad \phi_2 \text{ is } C\text{-Lipschitz on } \mathbf{R}^n$$

and that

$$(9.41) \qquad |\phi_2(z) - z| \leq C \quad \text{for all} \ z \in \mathbf{R}^n,$$

where C depends only on C_0 and n. As usual, we only need to worry about $h \circ \phi_2$, because of (9.22). Notice first that

$$h(\phi_2(x,y)) - y = \sum_{\xi \in \Xi} \lambda'_\xi(x,y)(y_\xi - y),$$

because

$$\lambda'_0(x,y) + \sum_{\xi \in \Xi} \lambda'_\xi(x,y) = 1$$

for all (x,y), by construction. (See (9.34)-(9.36).) On the other hand,

$$d_\xi(x,y) \geq 1 \quad \text{whenever} \quad |y - y_\xi| \geq C',$$

where C' is as in (9.29). This implies that

$$\lambda_\xi(x,y) = 0 \quad \text{when} \quad |y - y_\xi| \geq C',$$

by (9.32), and hence that

$$\lambda'_\xi(x,y) = 0 \quad \text{when} \quad |y - y_\xi| \geq C'.$$

From here we obtain

$$|h(\phi_2(x,y)) - y| \leq C' \quad \text{for all } x, y,$$

using also the fact that $\lambda'_\xi(x,y) \geq 0$ for all ξ and $\Sigma_\xi \lambda'_\xi(x,y) \leq 1$. Thus we get (9.41), with $C = C'$, since $\pi(\phi_2(x,y)) = x$. As for (9.40), it suffices to get a bound for the Lipschitz norm of $h(\phi_2(x,y)) - y$, and this follows from the above representation of this function as the sum of $\lambda'_\xi(x,y)(y_\xi - y)$ over ξ. Specifically, we already know that each $\lambda'_\xi(x,y)$ is Lipschitz with bounded norm, and that each $\lambda'_\xi(x,y)$ takes values in $[0,1]$ (by construction). For (x,y) fixed, we also have that $\lambda'_\xi(x,y) \neq 0$ for only a bounded number of ξ's, and that $|y_\xi - y| \leq C'$ for these ξ's. A simple computation (for differences of products) yields the desired bound for the Lipschitz norm of $h(\phi_2(x,y)) - y$. This proves (9.40).

Set

(9.42) $$E_2 = \phi_2(E_1).$$

Given $\xi \in \Xi$, let $\Gamma(\xi)$ denote the image of the d-plane $\{(x, y_\xi) : x \in P\}$ under the mapping ϕ_2. (As usual, $\xi = (x_\xi, y_\xi)$.) Since $\pi \circ \phi_2 = \pi$, we can write $\Gamma(\xi)$ more explicitly as

$$\Gamma(\xi) = \{(x, h \circ \phi_2(x, y_\xi)) : x \in P\}.$$

Thus $\Gamma(\xi)$ is a Lipschitz graph over P, and with bounded constant.

These graphs will be useful in describing E_2. Before we get to that, let us show that

(9.43) $$E_2 \cap \pi^{-1}(\text{int}(A_i)) = \emptyset \quad \text{when} \quad i \in I_1.$$

From (9.22) we have that $\pi(E_2) = \pi(E_1)$, and so we should check that

$$E_1 \cap \pi^{-1}(\text{int}(A_i)) = \emptyset \quad \text{when} \quad i \in I_1.$$

This can be derived from (9.19) and (9.17) in the case (9.18b). (Remember that $E_1 = \phi_1(E_0)$, by definition, as in (9.20).)

Next we want to verify that

(9.44) $$E_2 \cap \pi^{-1}(\text{int}(A_i)) \subseteq \Gamma(\xi_i) \quad \text{when} \quad i \in I_2,$$

where ξ_i denotes the point (x_i, y_i) provided by (9.5). Remember that
$$\xi_i \in \Xi,$$
by the convention made just after (9.23). Let $i \in I_2$ and
$$(x, y) \in E_2 \cap \pi^{-1}\big(\text{int}(A_i)\big)$$
be given, and let $(x_0, y_0) \in E_0$ be such that $(x, y) = \phi_2 \circ \phi_1(x_0, y_0)$. Set
$$(x_1, y_1) = \phi_1(x_0, y_0).$$
Notice that $x = x_1$, because of (9.22). This implies that x_1 lies in int(A_i). From here we may conclude that x_0 lies in A_i, and even in int(A_i), since $x_1 = \pi\big(\phi_1(x_0, y_0)\big)$, and using (9.19) and (9.17) in the case of (9.18a). (That is, we employ (9.19) for the other i's besides the given one, in order to eliminate the possibility that x_0 lies in some other A_j. Also, (9.16) ensures that $x_0 \in P_0$.) In addition, we have that $|y_0 - y_i| < 2$, (where (x_i, y_i) is again as in (9.5)), because otherwise $x_1 = \pi\big(\phi_1(x_0, y_0)\big)$ would lie in ∂A_i (rather than int(A_i)), by (9.17) in the case (9.18c). (Recall that θ_i takes values in ∂A_i, as indicated by its definition shortly before (9.14).) From (9.15) we have that $y_0 = y_1$, and hence $|y_1 - y_i| < 2$. On the other hand, $|x_1 - x_i| < d$, since x_1 and x_i both lie in A_i (which is a d-dimensional cube of unit size). Thus we conclude that
$$|x_1 - x_i| + |y_1 - y_i| < d + 2,$$
which means that (x_1, y_1) lies in $H(\xi_i)$, by (9.24). This permits us to apply (9.38) in order to conclude that $(x, y) = \phi_2(x_1, y_1)$ lies in $\Gamma(\xi_i)$. This completes the proof of (9.44).

We also want to control the sets $E_2 \cap \pi^{-1}(\partial A_i)$, $i \in I_0$. Let $i \in I_0$ and
$$(x, y) \in E_2 \cap \pi^{-1}(\partial A_i)$$
be given, let $(x_0, y_0) \in E_0$ be such that $(x, y) = \phi_2 \circ \phi_1(x_0, y_0)$, and set
$$(x_1, y_1) = \phi_1(x_0, y_0).$$
Thus $x_1 = x$, by (9.22), so that $x_1 \in \partial A_i$. This implies that x_0 lies in a cube A_j, $j \in I_0$, which intersects A_i, because of (9.19). (Notice that x_0 has to lie in some A_j with $j \in I_0$ – as opposed to $P \setminus P_0$ – simply because $(x_0, y_0) \in E_0$ and $\pi(E_0) \subseteq P_0$. The latter comes from (9.1) and the statement immediately following it. Also, $j = i$ is possible here.) Let ξ be an element of $\Xi(j)$ such that $|\xi - (x_0, y_0)| \leq 1$, as in (9.3). As before, we have that $y_1 = y_0$, by (9.15), and $|x_1 - x_0| < d$, by (9.19). (That is, x_0 and x_1 both lie in A_j, by (9.19), and A_j is a d-dimensional cube of unit size.) Hence $|x_1 - x_\xi| + |y_1 - y_\xi| \leq d + 2$, and $(x_1, y_1) \in H(\xi)$, by (9.24). Therefore, $(x, y) = \phi_2(x_1, y_1)$ lies on $\Gamma(\xi)$, by (9.38). This proves that

(9.45) $$E_2 \cap \pi^{-1}(\partial A_i) \subseteq \bigcup_{\xi \in \Xi} \Gamma(\xi)$$

for every $i \in I_0$. We can also restrict the union on the right-hand side to those ξ's that lie in $\Xi(j)$ for some $j \in I_0$ such that $A_j \cap A_i \neq \emptyset$.

We should emphasize that (9.43), (9.44), and (9.45) account for all of E_2, because
$$\pi(E_2) \subseteq P_0 = \bigcup_{i \in I_0} A_i.$$

9.1. A GENERAL DEFORMATION RESULT

Indeed, $\pi(E_0) \subseteq P_0$, because of our assumptions at the beginning of the section (i.e., (9.1) and the statement just after it). We also have that

$$\pi(E_1) = \pi(\phi_1(E_0)) \subseteq P_0,$$

using (9.19). This implies that $\pi(E_2) \subseteq P_0$, since $E_2 = \phi_2(E_1)$ and $\pi \circ \phi_2 = \pi$, by (9.22).

Set

(9.46) $$\Gamma = \bigcup_{i \in I_2} \left(\Gamma(\xi_i) \cap \pi^{-1}(A_i) \right).$$

Here ξ_i again denotes the point (x_i, y_i) provided by (9.5). From (9.43) and (9.44) we have that

$$E_2 \backslash \Gamma \subseteq \bigcup_{i \in I_0} \left(E_2 \cap \pi^{-1}(\partial A_i) \right).$$

This implies that

(9.47) $$H^{d-1}(E_2 \backslash \Gamma) < \infty,$$

because of (9.45). More precisely, (9.45) ensures that $E_2 \backslash \Gamma$ is contained in the finite union of Lipschitz graphs over $(d-1)$-dimensional cubes, the latter coming from the boundaries ∂A_i, $i \in I_0$.

Define Φ_t for $\frac{1}{3} \leq t \leq \frac{2}{3}$ by

(9.48) $$\Phi_t(z) = (2 - 3t)\phi_1(z) + (3t - 1)\phi_2 \circ \phi_1(z),$$

$z \in \mathbf{R}^n$. Thus $\Phi_{1/3} = \phi_1$ and $\Phi_{2/3} = \phi_2$. The set $E_2 = \Phi_{2/3}(E_0)$ is already much better than E_0 in terms of proving the main estimate (9.12), but we still need to eliminate most of the points in the graphs

$$\Gamma_i = \Gamma(\xi_i) \cap \pi^{-1}(A_i), \ i \in I_2,$$

of which Γ is composed.

The remainder of the construction. If $I_2 = \emptyset$, then in fact we are content with $\Phi_{2/3}$, and we can set $\Phi_t(z) = \Phi_{2/3}(z)$ for $t \geq \frac{2}{3}$. Otherwise we want to continue Φ_t with a new deformation ψ_u, $0 \leq u \leq 1$, which will remove nearly all of *one* of the Γ_i's. Afterwards we shall apply additional deformations, as needed, to get rid of the other Γ_i's.

We know from (9.4) that I_1 is not empty. If I_2 is not empty as well, then we can find cubes A_i, $i \in I_2$, and A_j, $j \in I_1$, which are contiguous, in the sense that A_i and A_j share a common $(d-1)$-dimensional face. Under these conditions, we claim that there is a continuous family of mappings ψ_u, $0 \leq u \leq 1$, on $\mathbf{R}^n$ with the following properties:

(9.49) $(u, z) \longrightarrow \psi_u(z)$ is a C-Lipschitz mapping from
$[0, 1] \times \mathbf{R}^n$ to $\mathbf{R}^n$;

(9.50) $\psi_u(z) = z$ for all z when $u = 0$, and for all u when
$\pi(z) \in P \backslash (A_i \cup A_j)$;

(9.51) if $\psi_u(z) \neq z$ for some $u \in [0, 1]$, then $\text{dist}(z, \Gamma_i) \leq 1$, and
$\psi_{u'}(z) \in \{w \in \pi^{-1}(A_i \cup A_j) : \text{dist}(w, \Gamma_i) \leq 1\}$ for all $u' \in [0, 1]$;

(9.52) $\psi_1(\Gamma_i) \subseteq \Gamma_i \cap \pi^{-1}(\partial(A_i \cup A_j))$.

The last property implies that we are making "progress", in the sense that the deformation ψ_u, $0 \leq u \leq 1$, has the effect of sweeping out the inside of Γ_i to the boundary of Γ_i by the end (when $u = 1$). At the same time, (9.50) ensures that none of the other Γ_ℓ's, $\ell \in I_2 \setminus \{i\}$, are disturbed.

The existence of this deformation ψ_u, $0 \leq u \leq 1$, is not hard to establish. Here is a basic recipe. The first step is to make a Lipschitz deformation $(u, x) \longrightarrow \widetilde{\psi}_u(x)$ from $[0,1] \times (A_i \cup A_j)$ to $A_i \cup A_j$ such that

$$\widetilde{\psi}_u(x) = x$$

for all x when $u = 0$, and for all u when x lies in $\partial(A_i \cup A_j)$ or $\mathrm{dist}(x, A_i) \geq C^{-1}$ (where C is taken to be somewhat larger than the Lipschitz constant for $\Gamma(\xi_i)$), and so that

$$\widetilde{\psi}_1(A_i) \subseteq \partial(A_i \cup A_j) \cap A_i.$$

(Note that $\partial(A_i \cup A_j)$ does not contain the $(d-1)$-dimensional face common to ∂A_i and ∂A_j.) This is easy to arrange; if one thinks of A_i as a box, then one makes it into an empty box by pushing in the $(d-1)$-dimensional face in $\partial A_i \cap \partial A_j$, while leaving the rest of ∂A_i alone. As a deformation on $A_i \cup A_j$ one cannot avoid displacing some points in the interior of A_j, but we can keep fixed the elements of ∂A_j that do not lie in the common face with A_i. We can also keep fixed the elements of A_j that do not lie with C^{-1} of A_i, and we can do these things in such a way that $\widetilde{\psi}_u(x)$ has bounded Lipschitz norm (as a function of both x and u).

We then define ψ_u, as a mapping from $\Gamma(\xi_i) \cap \pi^{-1}(A_i \cup A_j)$ to itself, through the condition

$$\pi(\psi_u(z)) = \widetilde{\psi}_u(\pi(z)).$$

This makes sense, because $\Gamma(\xi_i)$ is a graph over P. (Remember that

$$\Gamma_i = \Gamma(\xi_i) \cap \pi^{-1}(A_i)$$

(as in the notation just below (9.48)), but that $\Gamma(\xi_i)$ extends over all of P. The definition of $\Gamma(\xi)$ was given just after (9.42).) As a function on

$$[0,1] \times \big(\Gamma(\xi_i) \cap \pi^{-1}(A_i \cup A_j)\big),$$

$\psi_u(z)$ is Lipschitz with bounded constant, because of the bound for the Lipschitz norm of $\Gamma(\xi_i)$, and the one for the Lipschitz norm of $\widetilde{\psi}_u$. It is easy to check that $\psi_u(z)$ also satisfies (9.50)–(9.52) on $[0,1] \times \big(\Gamma(\xi_i) \cap \pi^{-1}(A_i \cup A_j)\big)$, to the extent that they are applicable, by construction. Once we have this, it is not hard to extend $\psi_u(z)$ to a mapping from $[0,1] \times \mathbf{R}^n$ into $\mathbf{R}^n$ that satisfies (9.49)–(9.52). (We omit the details.)

The set $\psi_1(E_2)$ is an improvement over E_2, in the sense that

(9.53) $$H^{d-1}\bigg(\psi_1(E_2) \setminus \bigcup_{\ell \in I_2 \setminus \{i\}} \Gamma_\ell\bigg) < \infty.$$

To see this, we start with the analogous statement (9.47) for E_2. This implies that

$$H^{d-1}\bigg(\psi_1\bigg(E_2 \setminus \bigcup_{k \in I_2} \Gamma_k\bigg)\bigg) < \infty,$$

using the fact that ψ_1 is Lipschitz. On the other hand, $\psi_1(\Gamma_\ell) = \Gamma_\ell$ when $\ell \in I_2 \setminus \{i\}$, because of (9.50), while $H^{d-1}(\psi_1(\Gamma_i)) < \infty$ by (9.52) (and because Γ_i is a Lipschitz graph). Thus we obtain (9.53).

If I_2 contains nothing more than i, then we are finished. That is, we can extend Φ_t to $0 \leq t \leq 1$ by setting

$$\Phi_t(z) = \psi_{3t-2} \circ \phi_2 \circ \phi_1(z) \quad \text{for } \tfrac{2}{3} \leq t \leq 1.$$

If I_2 contains at least one other element besides i, then we repeat the process. That is, we select $i_1 \in I_2 \setminus \{i\}$ and $j_1 \in I_1 \cup \{i\}$ such that A_{i_1} is contiguous to A_{j_1}, and we choose a deformation $\psi_v^{(1)}$, $0 \leq v \leq 1$, with properties like the ones enjoyed by ψ_u before. In particular, we want $\psi_u^{(1)}$ to satisfy

$$\psi_1^{(1)}(\Gamma_{i_1}) \subseteq \Gamma_{i_1} \cap \pi^{-1}\bigl(\partial(A_{i_1} \cup A_{j_1})\bigr).$$

We can then think of extending ψ_u to $u \in [0, 2]$ by setting

$$\psi_u(z) = \psi_{u-1}^{(1)}(\psi_1(z)) \quad \text{when } 1 \leq u \leq 2.$$

With this definition we have that

$$H^{d-1}\left(\psi_2(E_2) \setminus \bigcup_{\ell \in I_2 \setminus \{i, i_1\}} \Gamma_\ell\right) < \infty,$$

for the same reasons as in (9.53).

If i and i_1 are the only elements of I_2, then we stop here. Otherwise we continue with a third deformation $\psi_u^{(2)}$, $0 \leq u \leq 1$, like the ones before. If m denotes the total number of elements of I_2, then in the end we get a Lipschitz deformation ψ_u, $0 \leq u \leq m$, of $\mathbf{R}^n$. The key feature of this deformation is that if $E_3 = \psi_m(E_2)$, then

(9.54) $$H^{d-1}(E_3) < +\infty.$$

We define $\Phi_t(z)$ for $t \in [\tfrac{2}{3}, 1]$ by

(9.55) $$\Phi_t(z) = \psi_{3m\left(t-\tfrac{2}{3}\right)}\bigl(\phi_2(\phi_1(z))\bigr).$$

The rest of the proof of Proposition 9.6. Let us check that $\{\Phi_t\}_{0 \leq t \leq 1}$ enjoys all the properties required in Proposition 9.6. The Lipschitz condition for $(t, z) \mapsto \Phi_t(z)$ (with no specific bound) follows from the Lipschitz behavior of the individual pieces used to construct Φ_t, $0 \leq t \leq 1$, and from the continuity of Φ_t at the junctures where the pieces were combined, i.e., $t = \tfrac{1}{3}, \tfrac{2}{3}$, and $\tfrac{2}{3} + \tfrac{k}{3m}$, $k = 1, 2, \ldots, m$. This takes care of (9.7), and (9.8) is immediate.

If z is a point in $\mathbf{R}^n$ with $\operatorname{dist}(z, Q_0) \geq d + 3$, then $\phi_1(z) = z$ by (9.16), and then $\phi_2(z) = z$ by (9.39). Thus $\Phi_t(z) = z$ for $0 \leq t \leq \tfrac{2}{3}$.

To deal with $t > \tfrac{2}{3}$, let us check that the sets Γ_i (defined shortly after (9.48)) are contained in Q_0 for all $i \in I_2$. Remember from (9.1) that $Q_0 = P_0 \times V_0$, and that $\pi(\Gamma_i) = A_i \subseteq P_0$, by construction. On the other hand, Γ_i is contained in the graph $\Gamma(\xi_i)$ of $x \mapsto h(\phi_2(x, y_i))$, and $h(\phi_2(x, y_i))$ lies in V_0 for all x in P, simply because $h(\phi_2(x, y_i))$ is a convex combination of $y_i = y_{\xi_i}$ and y_ξ, $\xi \in \Xi$, by (9.37). (That the right-hand side of (9.37) is a *convex* combination comes from the basic properties of the coefficient functions (9.32)–(9.36). Remember also that the points

$\xi \in \Xi$ lie in E_0, as in (9.5). Hence they lie in Q_0, so that the y_ξ's all lie in V_0. The set V_0 was required to be convex from the beginning, just before (9.1).) Thus

$$\Gamma_i \subseteq Q_0$$

for all $i \in I_2$.

If $z \in \mathbf{R}^n$ satisfies $\operatorname{dist}(z, Q_0) \geq d + 3$, then

$$\operatorname{dist}(z, \Gamma_i) \geq \operatorname{dist}(z, Q_0) \geq d + 3$$

for each $i \in I_2$, and (9.51) implies that z is not moved by the deformation ψ_u, $0 \leq u \leq 1$, or similarly by any of its subsequent counterparts $\psi_u^{(\ell)}$, $1 \leq \ell \leq m - 1$. This proves that $\Phi_t(z) = z$ for all $t \in [0, 1]$, so that (9.9) holds.

Next we want to verify (9.10). Fix $z \in \mathbf{R}^n$, and let A denote the set of $t \in (0, 1]$ such that $\Phi_t(z)$ is "stable" for a short time before t, i.e., so that there is an $\epsilon > 0$ such that $\Phi_s(z) = \Phi_t(z)$ when $s \in [t - \epsilon, t]$. Let us check that if $t_0 \in (0, 1] \backslash A$, then

$$\operatorname{dist}(\Phi_{t_0}(z), E_0) \leq C$$

for some constant C (that depends only on n and our original constant C_0).

If $t_0 \leq \frac{1}{3}$, then the fact that $t_0 \notin A$ implies that $\phi_1(z) \neq z$, because of the definition (9.21) of Φ_t when $0 \leq t \leq \frac{1}{3}$. In this case we have that $\operatorname{dist}(z, E_0) < 1$, by (9.16), and $|\phi_1(z) - z| \leq d$, because of (9.15) and (9.19). Using (9.21) again we get that

$$\operatorname{dist}(\Phi_t(z), E_0) \leq d + 1$$

for all $t \in [0, \frac{1}{3}]$, and in particular for $t = t_0$.

Now suppose that $\frac{1}{3} < t_0 \leq \frac{2}{3}$. Then the definition (9.48) of Φ_t when $\frac{1}{3} \leq t \leq \frac{2}{3}$ implies that $\phi_2(\phi_1(z)) \neq \phi_1(z)$. (Otherwise $\Phi_t(z)$ is constant on $[\frac{1}{3}, \frac{2}{3}]$.) From (9.39) we obtain that $\operatorname{dist}(\phi_1(z), E_0) < d + 3$. This implies that

$$\operatorname{dist}(\Phi_t(z), E_0) \leq C \qquad \text{for } \tfrac{1}{3} \leq t \leq \tfrac{2}{3},$$

because of (9.41) and (9.48).

We are left with the case where $\frac{2}{3} < t_0 \leq 1$. Choose an integer k, $1 \leq k \leq m$, so that $\frac{2}{3} + \frac{(k-1)}{3m} < t_0 \leq \frac{2}{3} + \frac{k}{m}$ (where m is as in (9.55)). In this case there exist $i \in I_2$ and $j \in I_0$ such that the cubes A_i and A_j in P are adjacent, and

$$\Phi_t(z) \in \{w \in \pi^{-1}(A_i \cup A_j) : \operatorname{dist}(w, \Gamma_i) \leq 1\}$$

for all $t \in \left[\tfrac{2}{3} + \tfrac{(k-1)}{m}, \tfrac{2}{3} + \tfrac{k}{m}\right]$.

This comes from (9.51) and its analogues for the ("later") deformations $\psi_u^{(j)}$, $1 \leq j \leq m - 1$. Next, observe that $\Gamma_i = \Gamma(\xi_i) \cap \pi^{-1}(A_i)$ has bounded diameter, because $\Gamma(\xi_i)$ has bounded constant as a Lipschitz graph over P, and because the diameter of A_i is bounded (by $\sqrt{d}$). (Remember that $\Gamma(\xi_i)$ was defined just after (9.42), and that ϕ_2 is Lipschitz with bounded constant, as in (9.40).) We also have that $\phi_2(\xi_i)$ lies in Γ_i, i.e., $\phi_2(\xi_i) \in \Gamma(\xi_i)$ by the definition of $\Gamma(\xi_i)$, and $\pi(\xi_i) \in A_i$ automatically, since ξ_i denotes the point (x_i, y_i) given in (9.5). Combining these two pieces of information, we obtain that every element of Γ_i lies within a bounded distance from $\phi_2(\xi_i)$. We also know from (9.41) that $|\phi_2(\xi_i) - \xi_i|$ is bounded, and so every element of Γ_i lies within a bounded distance of ξ_i. Since $\xi_i \in E_0$ (as in

(9.5)), it follows that every element of Γ_i lies within a bounded distance of E_0. Combining this with the restriction on $\Phi_t(z)$ given above, we conclude that

$$\mathrm{dist}\big(\Phi_t(z), E_0\big) \leq C \quad \text{when} \quad \tfrac{2}{3} + \tfrac{k-1}{m} \leq t \leq \tfrac{2}{3} + \tfrac{k}{m}.$$

This completes the proof of our earlier claim, i.e., that

$$\mathrm{dist}\big(\Phi_{t_0}(z), E_0\big) \leq C$$

(for some constant C) when t_0 lies in $(0,1]\backslash A$. It is not hard to derive (9.10) from this, so that $\mathrm{dist}\big(\Phi_t(z), E_0\big) \leq C$ for all $t \in [0,1]$ (including $t = 0$, in which case $\Phi_t(z) = z$) as soon as $\Phi_s(z) \neq z$ for a single $s \in [0,1]$. Indeed, the latter condition implies that $(0,1]\backslash A$ is not empty (as one can check), and the arguments above provide the required bound when t lies in $(0,1]\backslash A$. The bound for $t \in (0,1]\backslash A$ implies the same bound for t in the closure of $(0,1]\backslash A$, by continuity. This leads to the same bound for all $t \in [0,1]$, because $\Phi_t(z)$ is constant on the intervals that make up the interior of A, by the definition of A and standard arguments.

Alternatively, one could use more completely the information gathered above, about the behavior of $\Phi_t(z)$ when t lies in each of the intervals

$$[0, \tfrac{1}{3}],\ [\tfrac{1}{3}, \tfrac{2}{3}],\ [\tfrac{2}{3} + \tfrac{k-1}{m}, \tfrac{2}{3} + \tfrac{k}{m}],\ 1 \leq k \leq m$$

(i.e., about what happens when $\Phi_t(z)$ is not constant on one of these intervals). This permits one to work in a finite total number of steps.

Now let us check (9.11). Of course $E_0 \subseteq Q_0$, because of our original assumptions (i.e., just after (9.1)). It is easy to see that $\phi_1(Q_0) \subseteq Q_0$, using (9.15) and (9.19). This implies that $\Phi_t(Q_0) \subseteq Q_0$ when $0 \leq t \leq \tfrac{1}{3}$, by (9.21) and the fact that Q_0 is convex (since P_0 and V_0 are). We also have that

$$\phi_2(Q_0) \subseteq Q_0.$$

Indeed, if $(x, y) \in Q_0$, then $\pi\big(\phi_2(x, y)\big) = x$, by (9.22), and so we only have to show that $h\big(\phi_2(x, y)\big)$ lies in V_0. From (9.37) we know that $h\big(\phi_2(x, y)\big)$ is a convex combination of y and the points y_ξ, $\xi \in \Xi$, and the latter lie in V_0 by construction. More precisely, the linear combination on the right side of (9.37) is a convex one because of the definitions and discussion of the coefficients in (9.33)–(9.36), and each y_ξ, $\xi \in \Xi$, lies in V_0 because (x_ξ, y_ξ), $\xi \in \Xi$, lies in $E_0 \subseteq Q_0$ by construction. (See (9.23) and (9.3).) This implies that $h\big(\phi_2(x, y)\big) \in V_0$, since V_0 is convex by assumption. (This was mentioned just before (9.1).) Thus $\phi_2(Q_0) \subseteq Q_0$, as desired. From here it follows that $\Phi_t(Q_0) \subseteq Q_0$ for all $t \in [\tfrac{1}{3}, \tfrac{2}{3}]$, because of (9.48), the convexity of Q_0, and the fact that $\phi_1(Q_0) \subseteq Q_0$. Combining this with the earlier observation for $t \in [0, \tfrac{1}{3}]$, we obtain that $\Phi_t(Q_0) \subseteq Q_0$ when $t \in [0, \tfrac{2}{3}]$.

Now consider the deformation ψ_u, $0 \leq u \leq 1$, from (9.49)–(9.52). Again we would like to have that $\psi_u(Q_0) \subseteq Q_0$ for all $u \in [0,1]$. This is not hard to arrange from the construction. To see this, it is helpful to consider $\pi \circ \psi_u$ and $h \circ \psi_u$ separately. For $\pi \circ \psi_u$, we have that $\pi\big(\psi_u(x, y)\big) = x$ when $x \notin A_i \cup A_j$, by (9.50), and it is easy to make sure that $\pi\big(\psi_u(x, y)\big) \in A_i \cup A_j$ when $x \in A_i \cup A_j$. Even if ψ_u were not constructed in this way originally, it could easily be "corrected" afterwards, using a Lipschitz retraction of P onto $A_i \cup A_j$ (which is a rectangular box), and this "correction" does not need to disturb (9.49)–(9.52). As for $h \circ \psi_u$, let us begin by noticing that $\Gamma(\xi) \cap \pi^{-1}(P_0) \subseteq Q_0$ for every ξ in Ξ. This follows from the definition of $\Gamma(\xi)$ as the image of the $y = y_\xi$ d-plane under ϕ_2 (discussed just after (9.42)), and the fact that $\pi \circ \phi_2 = \pi$ and $\phi_2(Q_0) \subseteq Q_0$ (as mentioned above). In particular, we have that

$$\Gamma(\xi_i) \cap \pi^{-1}(A_i \cup A_j) \subseteq Q_0.$$

This ensures that

$$\psi_u(z) \in Q_0 \quad \text{when } 0 \leq u \leq 1 \text{ and } z \in \Gamma(\xi_i) \cap \pi^{-1}(A_i \cup A_j),$$

because of the construction for ψ_u that we gave before (between (9.52) and (9.53)). Thus $h \circ \psi_u(z)$ lies in V_0 when $0 \leq u \leq 1$ and $z \in \Gamma(\xi_i) \cap \pi^{-1}(A_i \cup A_j)$. We want to say that $h \circ \psi_u(z)$ lies in V_0 for all $0 \leq u \leq 1$ and all $z \in Q_0$. This comes down to the choice of the Lipschitz extension made at the end of the construction of ψ_u, an extension that is supposed to be compatible with (9.50), (9.51), and with the definition of ψ_u on $[0,1] \times \big(\Gamma(\xi_i) \times \pi^{-1}(A_i \cup A_j)\big)$. The (new) requirement that $h \circ \psi_u(z) \in V_0$ when $0 \leq u \leq 1$ and $z \in Q_0$ is not too difficult to accommodate in these circumstances, and we omit the details. (One can think of making the extension in two steps; first to $z \in Q_0$, where the requirement that $h \circ \psi_u(z)$ lies in V_0 is easily achieved (or corrected after the fact if one prefers, e.g., through a Lipschitz retraction onto V_0), and then to z's outside Q_0. For the first step, the existence of a Lipschitz retraction from $\mathbf{R}^n$ onto V_0 is not hard to derive from the convexity of V_0, and it is especially obvious when V_0 is a cube or a ball.)

The bottom line is that the earlier construction of ψ_u, $0 \leq u \leq 1$, can be refined or corrected to give $\psi_u(Q_0) \subseteq Q_0$ for all u in $[0,1]$, in addition to (9.49)–(9.52). We can do the same thing for the subsequent deformations $\psi_v^{(k)}$, $0 \leq v \leq 1$, $1 \leq k \leq m-1$, and in exactly the same way. In the end we obtain that

$$\Phi_t(Q_0) \subseteq Q_0 \quad \text{for all } t \in [0,1],$$

because of (9.55) and our earlier observations for ϕ_2 and Φ_t, $0 \leq t \leq \frac{2}{3}$. This finishes the proof of (9.11).

Property (9.12) is the same as (9.54), just in different notation, and so we are left with (9.13). From (9.16) we know that ϕ_1 is equal to the identity mapping on $\pi^{-1}(P \backslash P_0) = \mathbf{R}^n \backslash (P_0 \times V)$, and so the same is true for each Φ_t, $0 \leq t \leq \frac{1}{3}$, by (9.21). We have already seen that ϕ_2 is Lipschitz on all of $\mathbf{R}^n$, and with a controlled constant, as in (9.40). This implies that $\Phi_t(z)$ is Lipschitz with bounded constant on $[0, \frac{2}{3}] \times \pi^{-1}(P \backslash P_0)$, because of the definition (9.48) of $\Phi_t(z)$ for $\frac{1}{3} \leq t \leq \frac{2}{3}$ and the preceding observations for $0 \leq t \leq \frac{1}{3}$. We also have that

$$\Phi_t \text{ maps } \pi^{-1}(P \backslash P_0) \text{ into itself for } 0 \leq t \leq \frac{2}{3},$$

since Φ_t is equal to the identity on $\pi^{-1}(P \backslash P_0)$ when $0 \leq t \leq \frac{1}{3}$, ϕ_2 satisfies (9.22), and using (9.48) again. For $t > \frac{2}{3}$ we have that

$$\Phi_t(z) = \Phi_{2/3}(z) \quad \text{when } z \in \pi^{-1}(P \backslash P_0),$$

because of (9.50) and its analogues for the subsequent deformations $\psi_v^{(k)}$. Thus the Lipschitz bound for $\Phi_t(z)$, $0 \leq t \leq \frac{2}{3}$, $z \in \pi^{-1}(P \backslash P_0)$, extends automatically to $0 \leq t \leq 1$, which implies (9.13).

This completes the proof of Proposition 9.6, since we have now shown that the deformation Φ_t, $0 \leq t \leq 1$, satisfies all of (9.7)–(9.13).

9.2. Application to quasiminimizers

Let us now use Proposition 9.6 to prove Main Lemma 8.7. Let
$$S, \; B = \overline{B}(x_0, r_0), \; F, \; \{\Sigma_j\}_{j \geq 0}, \; P \text{ and } \pi$$
be given, as in the beginning of Chapter 8. In particular, we have the following conditions:

(9.56) $$x_0 \in S^*, \; 0 < r_0 < \delta, \; 2B \subseteq U;$$

F is a compact Ahlfors-regular set such that

(9.57) $$S^* \cap B \subseteq F \subseteq S^*;$$

$\{\Sigma_j\}$ is a collection of partitions of F that satisfy (7.2), (7.3), (7.4), and (7.10), with $R = r_0$ and $E = S^*$; P is a d-plane in $\mathbf{R}^n$; and π denotes the orthogonal projection onto P.

In Main Lemma 8.7 we are also given constants $\gamma > 0$ and $C_1 > 0$, and we want to find C_2 (large) and $\eta > 0$ (small) such that the following is true. Pick a point $z \in F$ and an integer $j \geq 0$, and assume that

(9.58) $$B(z, 2C_2 2^{-j} r_0) \subseteq \frac{1}{2} B.$$

Let $T = T_j(z)$ denote the union of the cubical sets Q in Σ_j which intersect the ball $B(z, C_2 2^{-j} r_0)$, and assume that

(9.59) $$H^d(\pi(T)) \geq \gamma H^d(T).$$

(Keep in mind that the elements of Σ_j are themselves of size $\approx 2^{-j} r_0$, as in (7.2).) Then (we want to show that) z lies in $G_j(C_1, C_2, \eta)$, the "good" set of points $x \in F$ that satisfy (8.4) or (8.5).

From now on γ and C_1 should be taken as fixed. Imagine that we have particular values of C_2 and η in play, and let us proceed by contradiction. Thus we assume that we have $z \in F$ and $j \geq 0$ such that (9.58) and (9.59) hold, and that neither of (8.4) or (8.5) are true. This means that

(9.60) $$\pi(T) \text{ does not contain (all of) } P \cap B(\pi(z), C_1 2^{-j} r_0)$$

and

(9.61) $$H^d(\pi(R)) \leq (1 + 2\eta) H^d(R) H^d(\pi(T)) H^d(T)^{-1}$$

for every cubical set $R \in \Sigma_j$ such that $R \subseteq T$.

We want to show that this leads to a contradiction, at least if C_2 is large enough and η is small enough (depending only on the constants mentioned at the end of Main Lemma 8.7). To do this, we shall use (9.58)–(9.61) and Proposition 9.6 to show that we can deform a significant portion of S^* onto a set of H^d-measure almost equal to zero, thereby contradicting the quasiminimality of S. First we need to set some additional notation and normalizations, to place ourselves into the setting of Proposition 9.6.

We already have a d-plane P (given as in Chapter 8), and, as in the beginning of this chapter, we take V to be an $(n-d)$-plane in $\mathbf{R}^n$ orthogonal to P. For simplicity, let us choose V so that it also passes through the origin. Let π and h denote the orthogonal projections of $\mathbf{R}^n$ onto P and V, respectively. We identify $\mathbf{R}^n$ with $P \times V$, and we shall often denote points in $\mathbf{R}^n$ by $w = (x, y)$, where $x = \pi(w)$ lies in P and $y = h(w)$ lies in V, as before. (The requirement that V contain the origin in fact yields $w = \pi(w) + h(w)$, but the notation (x, y) is generally clearer for our purposes. We fix on P a system of orthonormal co-ordinates, so that it makes sense to speak of the origin and unit cube in P (with respect to these co-ordinates), or of cubes in P with sides parallel to the axes.)

It will be convenient to normalize distances by the requirement that

$$2C_1 2^{-j} r_0 = 1. \tag{9.62}$$

It is easy to reduce to this case by rescaling, since the problem at hand is invariant under dilations. We may also assume, without loss of generality, that

$$h(z) = 0 \quad \text{and} \quad \pi(z) \text{ is the center of the unit cube in } P, \tag{9.63}$$

by making a suitable translation, if necessary.

Because the problem only becomes more difficult when C_1 is increased, we may assume, without loss in generality, that C_1 is sufficiently large compared to the constants for the cubical patchwork $\{\Sigma_\ell\}_{\ell \geq 0}$ so that

$$\operatorname{diam} R \leq \frac{1}{10} \quad \text{for all} \quad R \in \Sigma_j. \tag{9.64}$$

(This also uses (9.62).)

Let N be a very large integer, to be chosen later. Denote by P_0 the cube in P with sidelength $2N$ which is centered at the origin and has sides parallel to the axes that we have fixed on P already. We shall take C_2 to be somewhat larger than N, in such a way that P_0 lies well inside $B(z, C_2 2^{-j} r_0)$.

Let A_i, $i \in I$, be the tiling of P by cubes of unit size (and vertices with integer co-ordinates), as discussed at the beginning of this chapter. This uses the system of orthonormal co-ordinates on P that we have fixed above. Let I_0 be the set of $i \in I$ such that $A_i \subseteq P_0$. Thus

$$P_0 = \bigcup_{i \in I_0} A_i.$$

In order to avoid confusion with these "true" cubes, we shall sometimes refer to the elements of Σ_j as pseudocubes.

Because of the normalizations (9.62) and (9.63), we obtain from (9.60) that

$$\pi(T) \text{ does not contain the interior of the unit cube in } P. \tag{9.65}$$

Our next (intermediate) goal will be to choose a radius ρ_0 such that

$$N \leq \rho_0 \leq CN$$

and $P_0 \times (\partial B(0, \rho_0) \cap V) \subseteq P \times V \cong \mathbf{R}^n$ lies sufficiently far away from F. Eventually, we shall apply Proposition 9.6 with $V_0 = V \cap B(x, \rho_0)$, and this choice of ρ_0 will help us to avoid the need for certain information about what our deformation does to S near the boundary of Q_0. Here and in the estimates below, C will be allowed to represent any constant that depends only on n, C_1, γ, the Ahlfors regularity constant for F, and the cubical patchwork constants for $\{\Sigma_\ell\}_\ell$, but not on N, C_2, or η.

LEMMA 9.66. *Let Σ denote the set of pseudocubes $R \in \Sigma_j$ such that $R \subseteq T$. Then*

$$H^d\bigl(\pi(R)\bigr) \geq (\gamma - CC_2^d \eta) H^d(R) \tag{9.67}$$

for all $R \in \Sigma$, *and*

(9.68) $$H^d\big(\pi(R_1) \cap \pi(R_2)\big) \leq CC_2^d \eta$$

for $R_1, R_2 \in \Sigma$ *such that* $R_1 \neq R_2$.

The point here is that we shall be able to take η as small as we want in (9.67) and (9.68), even depending on C_2. (This will not come until later in the section, however.)

We begin with (9.67). Set $\alpha = H^d\big(\pi(T)\big) H^d(T)^{-1}$, and fix $R_0 \in \Sigma$. Then

(9.69) $$H^d\big(\pi(T)\big) \leq \sum_{R \in \Sigma} H^d\big(\pi(R)\big)$$
$$\leq (1 + 2\eta)\alpha \sum_{R \neq R_0} H^d(R) + H^d\big(\pi(R_0)\big)$$

because of (9.61). Notice that

$$H^d\big(\pi(T)\big) = \alpha H^d(T) = \alpha \sum_{R \in \Sigma} H^d(R).$$

The first equality is just the definition of α, while the second relies on the fact that T is the disjoint union of the R's in Σ. This comes from the definition of T (just after (9.58)), which ensures that T contains all the pseudocubes in Σ_j that it meets, and it also uses the pairwise disjointness of distinct pseudocubes in Σ_j (see Definition 7.1). Subtracting $H^d\big(\pi(T)\big)$ from both sides of (9.69) we obtain that

(9.70) $$0 \leq -\alpha H^d(R_0) + 2\eta\alpha \sum_{R \neq R_0} H^d(R) + H^d\big(\pi(R_0)\big).$$

This yields

(9.71) $$H^d\big(\pi(R_0)\big) \geq \alpha H^d(R_0) - 2\eta\alpha \sum_{R \neq R_0} H^d(R)$$
$$\geq \alpha H^d(R_0) - 2\eta\alpha H^d(T)$$
$$\geq \gamma H^d(R_0)\left[1 - 2\eta H^d(T) H^d(R_0)^{-1}\right],$$

since $\alpha \geq \gamma$ by (9.59). The estimate (9.67) (for R_0) follows from this and a brutal estimate for $H^d(T) H^d(R_0)^{-1}$ using the Ahlfors-regularity of F and (7.2). (Remember also the precise definition of T, given just after (9.58).)

The proof of (9.68) is similar:

(9.72) $$H^d\big(\pi(T)\big) \leq \sum_{R \in \Sigma} H^d\big(\pi(R)\big) - H^d\big(\pi(R_1) \cap \pi(R_2)\big)$$
$$\leq (1 + 2\eta)\alpha \sum_{R \in \Sigma} H^d(R) - H^d\big(\pi(R_1) \cap \pi(R_2)\big)$$
$$= (1 + 2\eta) H^d\big(\pi(T)\big) - H^d\big(\pi(R_1) \cap \pi(R_2)\big)$$

again by (9.61) (for the second step). Subtracting $H^d\big(\pi(T)\big)$ as before yields

(9.73) $$H^d\big(\pi(R_1) \cap \pi(R_2)\big) \leq 2\eta H^d\big(\pi(T)\big)$$
$$\leq 2\eta H^d(T) \leq CC_2^d \eta,$$

with Ahlfors-regularity and the normalization (9.62) used in the last step (as well as the precise definition of T). This proves Lemma 9.66.

LEMMA 9.74. *There is a constant $C_3 > 10$, depending only on n, C_1, γ, the Ahlfors-regularity constant for F, and the cubical patchwork constants for $\{\Sigma_\ell\}$, so that the following is true. Assume that N is large enough and that $\eta > 0$ is small enough, where this may depend on the constants mentioned above, and also on C_2 in the case of η (but* not *in the case of N). If*

(9.75) $$2(C_3 + 10n)N \leq C_2 2^{-j} r_0,$$

then we can choose ρ_0 so that

(9.76) $$10N \leq \rho_0 \leq C_3 N$$

and

(9.77) $$\operatorname{dist}(P_0 \times (\partial B(0, \rho_0) \cap V), F) \geq N.$$

The hypothesis (9.75) will be useful because it ensures that all of the pseudocubes that will be considered in the proof lie in T. It will not cost us anything in the end, because we can take C_2 to be as large as we want.

In the following we shall sometimes write $B_V(0, r)$ instead of $B(0, r) \cap V$ i.e., for the ball inside of V. When we write $\partial B_V(0, r)$, we mean the boundary relative to V, or, equivalently, $\partial B(0, r) \cap V$.

To prove the lemma, suppose that none of the radii

$$\rho_0 = 10Nm, \ 1 \leq m \leq \frac{C_3}{10},$$

satisfy (9.77). We want to find a contradiction if C_3 is large enough and η is small enough.

Let j_0 denote the smallest integer such that all of the pseudocubes in Σ_{j_0} have diameter $\leq N$. Let us first check that j_0 is necessarily a *positive* integer, at least if C_3 is large enough. To see this, notice that the diameter of our set F is comparable to r_0. This is not hard to check, and it was pointed out before, in Chapter 8, between (8.1) and (8.2). Thus the diameters of the elements of Σ_{j_0} are comparable to $2^{-j_0} r_0$, by (7.2) in Definition 7.1. On the other hand, $C_3 N \leq r_0$, because of (9.75) and (9.58). (Remember that $B = \overline{B}(x_0, r_0)$ in (9.58).) This shows that j_0 must be positive, if C_3 is large enough. (Otherwise the diameters of the elements of Σ_{j_0} could not be less than or equal to N.)

Similarly, the *minimality* of j_0 implies that $2^{-j_0} r_0 \geq C^{-1} N$. From this last inequality we may conclude that

$$j_0 \geq j,$$

at least if N is large enough (depending on C_1 as well as geometric constants), because of (9.62). The lower bound for $2^{-j_0} r_0$ also implies that

$$H^d(R) \geq C^{-1} N^d$$

when $R \in \Sigma_{j_0}$, by (7.2).

For each value of m, $1 \leq m \leq \frac{C_3}{10}$, choose a pseudocube $R_m \in \Sigma_{j_0}$ such that

$$\operatorname{dist}(P_0 \times \partial B_V(0, 10Nm), R_m) < N.$$

We can do this, since we are assuming that $\rho_0 = 10Nm$ does not satisfy (9.77). The R_m's are disjoint, because they have diameter $\leq N$, by the definition of j_0,

9.2. APPLICATION TO QUASIMINIMIZERS

and because the different "shells" $P_0 \times \partial B_V(0, 10Nm)$ are at pairwise distance at least $10N$. It is not hard to verify that

$$R_m \subseteq T$$

for each m, using (9.75), the definition of T (just after (9.58)), and the definition of P_0 (shortly after (9.64)). The total mass of the R_m's can be estimated from below by

(9.78) $$H^d\left(\bigcup_m R_m\right) \geq C^{-1}C_3 N^d,$$

and their projections into P (by π) are contained in $\{x \in P : \mathrm{dist}(x, P_0) \leq 2N\}$. The latter implies that

(9.79) $$H^d\left(\pi\left(\bigcup_m R_m\right)\right) \leq (6N)^d.$$

We want to check that this contradicts Lemma 9.66 if C_3 and η are chosen correctly. Let Σ' denote the set of pseudocubes $Q \in \Sigma_j$ that are contained in $\bigcup_m R_m$. Keep in mind that $j_0 \geq j$, as verified earlier; in particular,

$$\bigcup_m R_m = \bigcup_{Q \in \Sigma'} Q.$$

Also, Σ' is contained in the class Σ defined in Lemma 9.66, since the R_m's are all contained in T, as mentioned before.

Let $\mathcal{A}$ denote the set of pairs (Q_1, Q_2) which satisfy $Q_1, Q_2 \in \Sigma'$ and $Q_1 \neq Q_2$. Then

(9.80) $$H^d\left(\pi\left(\bigcup_m R_m\right)\right) = H^d\left(\pi\left(\bigcup_{Q \in \Sigma'} Q\right)\right)$$
$$\geq \sum_{Q \in \Sigma'} H^d(\pi(Q)) - \sum_{(Q_1, Q_2) \in \mathcal{A}} H^d(\pi(Q_1) \cap \pi(Q_2)).$$

This inequality follows from very general considerations of characteristic functions of sets and their unions, and is quite brutal. To put it plainly, if a point x lies in $\ell \geq 1$ of the sets $\pi(Q)$, $Q \in \Sigma'$, then x is counted once in $H^d\left(\pi\left(\bigcup_{Q \in \Sigma'} Q\right)\right)$, ℓ times in the first sum on the right-hand side of (9.80), and $\geq \ell - 1$ times in the second sum in (9.80), which is being subtracted. We do not mind being a bit crude in this estimate, because of the tight control over the masses of the intersections that we have from Lemma 9.66, as we are about to see.

The first sum on the right-hand side of (9.80) is $\geq (\gamma - CC_2^d \eta) H^d(\bigcup_m R_m)$, by (9.67) and the fact that $\bigcup_m R_m = \bigcup_{Q \in \Sigma'} Q$. The second sum (over $\mathcal{A}$) in (9.80) is bounded from above by $CC_2^d \eta \cdot (\#\mathcal{A})$, where $\#\mathcal{A}$ denotes the number of elements of $\mathcal{A}$, because of (9.68). To estimate $\#\mathcal{A}$ we again proceed brutally, observing that $\#\mathcal{A} \leq (\#\Sigma')^2$ and that $\#\Sigma' \leq \#\Sigma$. For Σ we have the bound $\#\Sigma \leq C \cdot C_2^d$, which follows from the definition of Σ (in Lemma 9.66), the definition of T (just after (9.58)), and the usual condition (7.2) for the pseudocubes in $\Sigma_j \subseteq \Sigma$. (More precisely, (7.2) provides an approximation to the H^d-measure of every $Q \in \Sigma_j$, while the Ahlfors regularity of F gives an upper bound for $H^d(T)$. We automatically have $H^d(T) \geq \sum_{Q \in \Sigma'} H^d(Q)$, since $T \supseteq \bigcup_{Q \in \Sigma'} Q$ (and the Q's in $\Sigma' \subseteq \Sigma_j$ are pairwise disjoint, as in Definition 7.1), and this leads to the above bound for $\#\Sigma'$.) Thus

$\#\mathcal{A} \leq CC_2^{2d}$, and the sum over $\mathcal{A}$ in (9.80) is bounded from above by $CC_2^{3d}\eta$. If η is small enough (depending on C_2 in addition to γ and the geometric constants), then (9.80) yields

$$(9.81) \qquad H^d\Big(\pi\Big(\bigcup_m R_m\Big)\Big) \geq \frac{\gamma}{2} H^d\Big(\bigcup_m R_m\Big).$$

This also uses the lower bound (9.78) (in order to absorb the (small) upper bound for the sum over $\mathcal{A}$ into the $H^d\big(\bigcup_m R_m\big)$ term). If C_3 is large enough, then (9.81) is not compatible with (9.78) and (9.79). Thus we get a contradiction, and Lemma 9.74 follows.

Let us say a little more about the way that our constants will (eventually) be chosen. Right now we have three constants still "in the air", namely N, C_2, and η. We shall choose N near the end of the argument, depending on n, C_1, γ, the cubical patchwork constants for $\{\Sigma_\ell\}$, and the quasiminimizing constant k for our original set S. The only condition that we shall need to impose on C_2 is that it be large enough for (9.75) to hold. Thus we can think of C_2 as being determined now, modulo the choice of N. We shall soon say how η should be selected, depending on N and C_2, so that N will be the only free parameter remaining. (Of course it is important that we never try to choose N depending on η or C_2, since otherwise we would get caught in a vicious circle.)

With this in mind, let us now fix a ρ_0 (depending on N) as in Lemma 9.74, so that (9.76) and (9.77) hold. Put

$$(9.82) \qquad V_0 = V \cap \overline{B}(0, \rho_0), \quad Q_0 = P_0 \times V_0 \subseteq \mathbf{R}^n.$$

We want to apply Proposition 9.6 to the set $E_0 = F \cap Q_0$. The next lemma will help us to verify the hypotheses of Proposition 9.6 (as described at the beginning of this chapter).

LEMMA 9.83. *If $\eta > 0$ is small enough (depending on N and C_2), then*

$$(9.84) \qquad \textit{for each } i \in I, \textit{ there are at most } C \textit{ pseudocubes } R \in \Sigma$$
$$\textit{such that } \pi(\overline{R}) \cap A_i \neq \emptyset$$

and

$$(9.85) \qquad \textit{for each } i \in I, \textit{ we can find a point } x_i \in \tfrac{1}{2} A_i \textit{ which lies in}$$
$$\pi(\overline{R}) \textit{ for at most one } R \textit{ in } \Sigma.$$

Recall that the cubes A_i, $i \in I$, were discussed just before (9.65), and that Σ was defined in Lemma 9.66. (For later application it is useful to have I here, in Lemma 9.83, rather than just I_0.)

To prove (9.84), we use (9.67) and (9.68), as follows. If η is small enough, then (9.67) implies that

$$H^d\big(\pi(R)\big) \geq \tfrac{\gamma}{2} H^d(R)$$

for all $R \in \Sigma$. Since $\Sigma \subseteq \Sigma_j$, $H^d(R)$ is comparable to $(2^{-j}r_0)^d$, as in (7.2). (This also relies on the fact that our original Ahlfors-regular set F (of which the Σ_ℓ's are partitions) has $\text{diam } F \approx r_0$. This came up before, in the initial part of the proof of Lemma 9.74.) Using the normalization (9.62), we can reformulate this as saying that $H^d(R)$ is bounded from above and below by positive numbers that depend only on C_1 and geometric constants. In the end we conclude that $H^d\big(\pi(R)\big)$ is bounded

9.2. APPLICATION TO QUASIMINIMIZERS

from below by a positive quantity that depends only on γ, C_1, and geometric constants.

Let us now fix $i \in I$, and consider only the R's in Σ such that $\pi(\overline{R})$ intersects A_i. For these R's we have that $\pi(R) \subseteq 2A_i$, because of (9.64). (Keep in mind that the A_i's are of unit size.) If the various $\pi(R)$'s were disjoint, then we could easily get a bound on their number from these two pieces of information, i.e., inclusion in $2A_i$, together with a lower bound on their measures individually. As it is, they need not be disjoint, but (9.68) provides an adequate substitute for this, at least if η is small enough. Specifically, if η is sufficiently small (depending on C_2 in particular), then

$$H^d\big(\pi(R) \cap \pi(R')\big) \leq \frac{1}{2^{d+1}} H^d\big(\pi(R)\big) H^d\big(\pi(R')\big)$$

for all $R, R' \in \Sigma$ such that $R \neq R'$. This follows from (9.68) and the lower bounds for the $H^d\big(\pi(R)\big)$'s discussed above. If f denotes the sum of the characteristic functions of the $\pi(R)$'s with $R \in \Sigma$ and $\pi(R) \subseteq 2A_i$, then one can use the bound for $H^d\big(\pi(R) \cap \pi(R')\big)$ to get that

$$\int_{2A_i} f^2 \leq \int_{2A_i} f + \frac{1}{2^{d+1}} \Big(\int_{2A_i} f\Big)^2.$$

This simply comes from expanding f^2 out as a double sum. More precisely, the contribution of the diagonal $(R = R')$ part reduces to $\int_{2A_i} f$, since the square of a characteristic function is equal to itself, while the off-diagonal terms lead to expressions like $H^d\big(\pi(R) \cap \pi(R')\big)$. These can be replaced by $2^{-(d+1)} H^d\big(\pi(R)\big) H^d\big(\pi(R')\big)$ and regrouped to get the square of the integral of f on the right-hand side of the inequality.

On the other hand,

$$\Big(\int_{2A_i} f\Big)^2 \leq |2A_i| \int_{2A_i} f^2 = 2^d \int_{2A_i} f^2,$$

by the Cauchy-Schwarz inequality. Combining this with the earlier estimate we get that

$$\int_{2A_i} f^2 \leq \int_{2A_i} f + \frac{1}{2} \int_{2A_i} f^2,$$

and hence that

$$\int_{2A_i} f^2 \leq 2 \int_{2A_i} f.$$

(All of these integrals are automatically finite, because there are only finitely many R's. Thus there is no problem with the subtraction here.) Using the Cauchy-Schwarz inequality again, we get that

$$\int_{2A_i} f \leq 2 |2A_i|.$$

This is exactly what we wanted (for (9.84)), because the lower bound for $H^d\big(\pi(R)\big)$ explained before enables us to convert this into an upper bound for the total number of R's.

Let us emphasize that the bound that we get for (9.84) by the end depends on C_1, γ, and geometric constants (like d and the constants for the cubical patchwork $\{\Sigma_\ell\}$, but not on N or C_2. This is consistent with the general declaration made just before Lemma 9.66. Although we need η to be small in a way that depends on C_2 for the argument to work, the value of η is not involved in the constants that we have in the estimates above. (Specifically, we used η to get a factor of $2^{-(d+1)}$ at a certain moment, but then this number itself does not involve η.)

Now let us prove (9.85). We first recall that
$$H^d(\overline{R}\backslash R) = 0$$
for all $R \in \Sigma$, because of (7.4) (or (7.10)). Thus
$$H^d\big(\pi(\overline{R}\backslash R)\big) = 0,$$
for all $R \in \Sigma$.

Fix $i \in I$, and let x be a point in $\frac{1}{2}A_i$. Imagine that x lies in $\pi(\overline{R}_1) \cap \pi(\overline{R}_2)$ for some pair of distinct pseudocubes $R_1, R_2 \in \Sigma$. We know from (9.68) (and the fact that $H^d\big(\pi(\overline{R}_i\backslash R_i)\big) = 0$) that $\pi(\overline{R}_1) \cap \pi(\overline{R}_2)$ has H^d-measure $\leq CC_2^d\eta$. We also know that the number of possible pairs (R_1, R_2) which can arise in this manner is bounded, because of (9.84). (More precisely, if $\pi(\overline{R})$ intersects $\frac{1}{2}A_i$, then $\pi(R)$ itself must intersect A_i, which puts us in the situation of (9.84).) Thus, if we consider the total set of x's in $\frac{1}{2}A_i$ which lie in $\pi(\overline{R}_1) \cap \pi(\overline{R}_2)$ for at least one pair of distinct pseudocubes $R_1, R_2 \in \Sigma$, then the measure of this set is $\leq CC_2^d\eta$ (with a slightly larger constant C than before). If η is small enough, depending on C_2 and the other constants, then this bound for the measure of the "bad" set (of double points in $\frac{1}{2}A_i$) is strictly less than $H^d\big(\frac{1}{2}A_i\big)$. Thus we conclude that there is at least one point x_i in A_i which does not lie in $\pi(\overline{R}_1) \cap \pi(\overline{R}_2)$ for any pair R_1, R_2 of distinct cubes in Σ, as in (9.85). This completes the proof of Lemma 9.83.

We now choose η, depending on N and C_2, so small that the conclusions of Lemmas 9.66 and 9.83 hold. (In other words, we shall not impose any further conditions on η beyond this point.) Our last free parameter N will be chosen at the very end of the argument.

We want to apply Proposition 9.6 with

(9.86) $$E_0 = F \cap Q_0$$

(and with Q_0 chosen as in (9.82)). Notice first that

(9.87) $$Q_0 \subseteq B(z, C_2 2^{-j} r_0).$$

This follows from (9.82), the normalization (9.63), the definition of P_0 (just after (9.64)), the upper bound for ρ_0 in (9.76), and (9.75). (The factor of 2 on the left side of (9.75) is helpful in this regard.) The definition of T (just before (9.59)) now ensures that

(9.88) $$E_0 \subseteq T,$$

and indeed T contains any element of Σ_j that intersects E_0. (Keep in mind that the pseudocubes in Σ_j cover F, and in fact form a partition of F, because of the way that we chose Σ_j (as reviewed just after (9.57)).)

If Σ is as in Lemma 9.66, then

(9.89) $$E_0 \subseteq \bigcup_{R \in \Sigma} R,$$

since $E_0 \subseteq T$ (and since Σ_j covers F).

Let us check that the hypotheses of Proposition 9.6 are satisfied. For (9.3) we employ Lemma 9.83. Specifically, for each $i \in I_0$ we define a set $\Xi(i)$ by picking exactly one element of $E_0 \cap \pi^{-1}(A_i) \cap R$ for each $R \in \Sigma$ that intersects $E_0 \cap \pi^{-1}(A_i)$. With this choice, (9.3) follows from (9.84) and the fact that diam $R \leq \frac{1}{10}$ for all $R \in \Sigma$, by (9.64).

Next, (9.4) follows from (9.60) and the normalizations (9.62) and (9.63). More precisely, the index $i \in I_0$ for which A_i is the unit cube in P is an element of I_1.

Finally, we can derive (9.5) from (9.85). Given $i \in I_2$, we choose $x_i \in \frac{1}{2} A_i$ as in (9.85), and we select $y_i \in V_0$ so that $(x_i, y_i) \in E_0$. (Such y_i exists, because otherwise i would lie in I_1. Every other $y \in V$ with $(x_i, y) \in E_0$ satisfies $|y - y_i| \leq 1$, because (x_i, y) must lie in the same pseudocube R as (x_i, y_i), by (9.85), and because these pseudocubes have diameter $\leq \frac{1}{10}$, as in (9.64).)

Thus we may apply Proposition 9.6, to get a deformation $\{\Phi_t\}_{0 \leq t \leq 1}$ on $\mathbf{R}^n$ with the properties (9.7)–(9.13). Now we want to test the quasiminimality of S using the mapping $\phi = \Phi_1$.

Let us check that $\phi(S)$ is an acceptable competitor. The first condition (1.3) (i.e., that ϕ be Lipschitz) follows from (9.7). Set $W = \{w \in \mathbf{R}^n : \phi(w) \neq w\}$, as usual. Then

$$(9.90) \qquad W \subseteq \{w \in \mathbf{R}^n : \operatorname{dist}(w, Q_0) \leq d + 3\}$$

by (9.9), and

$$(9.91) \qquad \phi(W) \subseteq \{w \in \mathbf{R}^n : \operatorname{dist}(w, Q_0) \leq C\}$$

by (9.10). This also uses the fact that $E_0 \subseteq Q_0$ (as in (9.86)). From here we get that

$$(9.92) \qquad W \cup \phi(W) \subseteq B(z, nN + C_3 N + C) \subseteq \frac{1}{2} B,$$

where z is the point given to us at the beginning (just before (9.58)), and $B = \overline{B}(x_0, r_0)$ (as in the notation just prior to (9.56)). To see this, notice that

$$Q_0 \subseteq B(z, nN + C_3 N + d),$$

by the definition (9.82) of Q_0, (9.76), the definition of P_0 (just after (9.64)), and the fact that z is the center of the unit cube in P (as in (9.63)). This implies the first inclusion in (9.92), using also (9.90) and (9.91). The second inclusion in (9.92) follows from (9.58) and (9.75), at least if N is large enough (i.e., large enough so that $C \leq N$, for instance, where C is as in (9.92)).

Conditions (1.5) and (1.6) on ϕ (concerning the size of $W \cup \phi(W)$) follow from (9.92) and (9.56). The last condition (1.7), about the existence of a suitable homotopy from the identity mapping to ϕ, can be derived from Proposition 9.6 in exactly the same way as for conditions (1.5) and (1.6) for ϕ. One can also get (1.7) by using Remark 1.10 and (9.92).

This shows that $\phi(S)$ is an acceptable competitor for S. Thus we may apply (1.8) to obtain

$$(9.93) \qquad H^d(S \cap W) \leq k H^d(\phi(S \cap W)).$$

To make use of this, we observe first that

$$(9.94) \qquad S^* \cap W = F \cap W.$$

This holds because S^* and F coincide in B (by (9.57)) and $W \subseteq \frac{1}{2} B$ (as in (9.92)). Set

$$(9.95) \qquad Q_1 = \{w \in \mathbf{R}^n : \operatorname{dist}(w, Q_0) \leq d + 3\},$$

and let us verify that

(9.96) $$H^d(\phi(S\cap W)) = H^d(\phi(S^*\cap W))$$
$$\leq H^d(\phi(F\cap Q_1)) = H^d(\phi(F\cap (Q_1\setminus Q_0))).$$

For this we use $H^d(S\setminus S^*) = 0$ (as in (1.13)) in the first step, (9.94) and the inclusion $W \subseteq Q_1$ (from (9.90)) in the second step, and the fact that $\phi(F\cap Q_0) = \phi(E_0) = \Phi_1(E_0)$ has zero H^d-measure, by (9.12), in the third. (Remember that $E_0 = F\cap Q_0$, as in (9.86).)

Put $F_1 = F\cap (Q_1\setminus Q_0)$. Using (9.77) and the definitions (9.82), (9.95) of Q_0, Q_1, one can check that

$$F_1 \cap \pi^{-1}(P_0) = \emptyset,$$

at least if N is larger than $d+3$. Thus

(9.97) $$F_1 = F_1\setminus \pi^{-1}(P_0) = (F\cap Q_1)\setminus (Q_0 \cup \pi^{-1}(P_0))$$
$$= (F\cap Q_1)\setminus \pi^{-1}(P_0),$$

where the last step uses the fact that $Q_0 \subseteq \pi^{-1}(P_0)$, by the definition (9.82) of Q_0. We also know from (9.13) that $\phi = \Phi_1$ is C-Lipschitz on $\mathbf{R}^n\setminus \pi^{-1}(P_0)$, and on F_1 in particular. Thus (9.96) leads to

(9.98) $$H^d(\phi(S\cap W)) \leq H^d(\phi(F_1)) \leq CH^d(F_1).$$

To estimate $H^d(F_1)$, we first cover $\pi(Q_1)\setminus P_0$ by cubes A_i, $i\in I$. Note that $\pi(Q_0) = P_0$, by the definition (9.82) of Q_0, so that $\pi(Q_1)$ consists exactly of the points $x\in P$ such that $\mathrm{dist}(x, P_0) \leq d+3$, by the definition (9.95) of Q_1. From this it follows that $\pi(Q_1)\setminus P_0$ can be covered by $\leq C(d)N^{d-1}$ cubes A_i, $i\in I$, because P_0 is a cube of sidelength $2N$ and the A_i's have sidelength 1 and disjoint interiors.

Let us observe that

(9.99) $$Q_1 \subseteq B(z, C_2 2^{-j} r_0),$$

for the same reasons as for (9.87) before, at least if N is larger than $d+3$. Similarly,

(9.100) $$F_1 \subseteq T,$$

and F_1 is covered by the pseudocubes R in Σ, for the same reasons as for (9.88) and (9.89) (assuming that $N \geq d+3$).

If A_i is one of the cubes in P used to cover $\pi(Q_1)\setminus P_0$, then $\pi^{-1}(A_i) \cap F_1$ is contained in the union of at most C pseudocubes R in Σ, because of (9.84). This yields

(9.101) $$H^d(\pi^{-1}(A_i) \cap F_1) \leq C,$$

since F_1 is covered by the pseudocubes in Σ, as mentioned above, and since the individual pseudocubes have bounded H^d measure. (The latter was explained before, in the first part of the proof of Lemma 9.83. It comes down to (7.2), $\mathrm{diam}\, F \approx r_0$, and the normalization (9.62).) Summing over i we obtain that

(9.102) $$H^d(F_1) \leq CN^{d-1},$$

since there are only CN^{d-1} choices of i that are needed to cover F_1 (as above). We can combine this with (9.98) and (9.93) to conclude that

(9.103) $$H^d(S\cap W) \leq CkN^{d-1}.$$

We want to contradict (9.103) with a suitable lower bound for $H^d(S \cap W)$. From (9.12) we have that

(9.104) $$H^d(\phi(E_0)) = H^d(\Phi_1(E_0)) = 0,$$

since $\phi = \Phi_1$, by definition. This implies that H^d-almost every element of E_0 must lie in W, since W consists of the points which are not fixed by ϕ. Of course $E_0 \subseteq F \subseteq S$, by (9.86) and (9.57), and so we get

$$H^d(S \cap W) \geq H^d(E_0).$$

The main point now is that

(9.105) $$H^d(E_0) \geq C^{-1} N^d,$$

because of Ahlfors regularity. To see this, let us first check that

(9.106) $$E_0 \supseteq F \cap B(z, N-d).$$

Remember that z is the point z from the beginning of the story (just before (9.58)), and that $E_0 = F \cap Q_0$, as in (9.86). From (9.82) we have that Q_0 is the Cartesian product of P_0 and a ball in V of radius ρ_0 and center 0, where $\rho_0 \geq 10N$ by (9.76). By definition (shortly after (9.64)), P_0 is a cube of size $2N$ which is centered at the "origin" in P. Using the normalization (9.63), we may conclude that

(9.107) $$Q_0 \supseteq B(z, N-d).$$

(We take $N-d$ for the radius here, rather than N, in order to compensate for the distance from $\pi(z)$ to the "origin" in P, which is bounded by d, by (9.63). There is no problem like this in the V directions, also by (9.63).) This implies (9.106), since $E_0 = F \cap Q_0$.

To derive (9.105) from Ahlfors regularity, let us be a bit careful and check that the ball $B(z, N-d)$ is in the correct range. We know that $z \in F$, as stated just before (9.58), and so that is fine. We also have that

(9.108) $$B(z, N-d) \subseteq B(z, C_2 2^{-j} r_0) \subseteq \frac{1}{2} B.$$

The first inclusion follows from (9.87) (or (9.75)), while the second comes from (9.58). These two pieces of information about z and N enable us to obtain (9.105) from the local Ahlfors-regularity result Proposition 4.1 for S^*, using also (9.57) and (9.56) (to put us into the position of being able to apply Proposition 4.1). (One could also think in terms of Ahlfors-regularity for F itself, but in the end this comes back to S^* anyway.)

Thus we get (9.105). By choosing N large enough (depending only on suitable constants), we can get a contradiction with (9.103) (i.e., since we have N^d in (9.105) and N^{d-1} in (9.103)). This is what we wanted, and it finishes the proof of Main Lemma 8.7.

Since we already saw in Chapter 8 how Proposition 8.15 can be derived from Main Lemma 8.7, the proof of Proposition 8.15 is now complete as well.

CHAPTER 10

Big Projections

We are now ready to finish the proof of our main result, Theorem 2.11. The principal missing component is the following assertion concerning the existence of "big projections".

PROPOSITION 10.1. *Let S be a (U, k, δ)-quasiminimizer and let $B = B(x_0, r_0)$ be a ball centered on S^* such that $r_0 < \delta$ and $B \subseteq U$. Then there is a d-plane P in $\mathbf{R}^n$ such that*

(10.2) $$H^d\bigl(\pi(S^* \cap B)\bigr) \geq C_4^{-1} R^d,$$

where π denotes the orthogonal projection of $\mathbf{R}^n$ onto P, and where C_4 depends only on n and k.

Before we give the proof of Proposition 10.1, let us first explain how Theorem 2.11 can be derived from the proposition. Let S be a quasiminimizer, and let $B = B(x, R)$ be a ball centered on S^* which satisfies (2.12). Because of Proposition 4.1, the set $E = S^* \cap B(x, 2R)$ satisfies the local Ahlfors-regularity condition required in Proposition 7.6. Thus there is a compact Ahlfors-regular set F of dimension d such that $S^* \cap B \subseteq F \subseteq S^* \cap \bigl(\frac{3}{2}B\bigr)$, and a cubical patchwork $\{\Sigma_j\}_{j \geq 0}$ on F which is adapted to $S^* \cap B(x, 2R)$, as in Proposition 7.6. We want to show that F is uniformly rectifiable, and that it contains big pieces of Lipschitz graphs, with constants that depend only on k and n. (This is exactly what is needed for Theorem 2.11. See Chapter 2 for the definitions.)

Let us first prove that F is uniformly rectifiable. Let $x_0 \in F$ and $0 < r_0 < \operatorname{diam} F$ be given. The pair x_0, r_0 will have the same role for F as x, r have in Definition 2.3 (but we give them different names because of the notation already set in the previous paragraph). Choose $j \geq 0$ to be the smallest integer such that every $Q \in \Sigma_j$ has diameter $< r_0$, and take Q_0 to be the element of Σ_j that contains x_0. Notice that

$$Q_0 \subseteq F \cap B(x_0, r_0),$$

and that

$$\operatorname{diam} Q_0 \geq C^{-1} r_0,$$

because of the minimality of j, and the property (7.2) of the cubical patchwork $\{\Sigma_j\}$. Also, if τ is chosen small enough (depending only on k and n), then we can use (7.10) and (7.2) to find a point ξ in Q_0 that satisfies

$$\operatorname{dist}\bigl(\xi, [S^* \cap B(x, 2R)]\backslash Q_0\bigr) > \tau \operatorname{diam} Q_0.$$

(This is easy to check.) On the other hand, $F \subseteq B\left(x, \frac{3R}{2}\right)$, $\operatorname{diam} Q_0 \leq \operatorname{diam} F \leq 3R$, and hence

$$\operatorname{dist}\left(\xi, S^* \backslash B(x, 2R)\right) \geq \frac{R}{2} \geq \frac{1}{6} \operatorname{diam} Q_0,$$

since $\xi \in F$. Let us shrink τ a bit, if necessary, to ensure that $\tau < \frac{1}{6}$. This yields

(10.3) $$\operatorname{dist}\left(\xi, S^* \backslash Q_0\right) > \tau \operatorname{diam} Q_0.$$

Now we want to apply Proposition 8.15 to the ball $\overline{B}(\xi, \tau \operatorname{diam} Q_0)$. Notice that $\tau \operatorname{diam} Q_0 < \frac{R}{2} < \frac{1}{2} \operatorname{dist}(\xi, \mathbf{R}^n \backslash U)$, since $B(x, 3R) \subseteq U$, as in (2.12) (which is part of our hypothesis now). Thus the double of $\overline{B}(\xi, \tau \operatorname{diam} Q_0)$ is contained in U, as required in Proposition 8.15. The conclusion of Proposition 8.15 gives us a compact set Γ contained in $S^* \cap B(\xi, \tau \operatorname{diam} Q_0)$ such that

$$H^d(\Gamma) \geq C_3^{-1} (\tau \operatorname{diam} Q_0)^d \geq C^{-1} r_0^d$$

(by our choice of Q_0), and a C_3-bilipschitz mapping h from Γ to a subset of $\mathbf{R}^d$. Since $\Gamma \subseteq Q_0 \subseteq F \cap B(x_0, r_0)$, by (10.3) and the choice of Q_0, Γ is the big bilipschitz image of a piece of $\mathbf{R}^d$ which is required in Definition 2.3. This proves that F is uniformly rectifiable. (Note that we have not used Proposition 10.1 so far.)

Essentially the same argument permits one to show that F enjoys the property of big projections (Definition 2.9). One simply uses Proposition 10.1 in place of Proposition 8.15.

We may now apply a general result, to the effect that an Ahlfors-regular set which is uniformly rectifiable and has the property of big projections also contains big pieces of Lipschitz graphs (Definition 2.7). More precisely, any Ahlfors-regular set with big projections is known to have big pieces of Lipschitz graphs if it satisfies the "weak geometric lemma" (WGL), by Theorem 1.14 on p. 857 of [7]. It is also known that uniform rectifiability implies the WGL. (Uniform rectifiability is actually substantially stronger than the WGL by itself, by an example given in [6].) For an explicit statement of the fact that uniform rectifiability implies the WGL, one can use Theorem 1.57 on p. 22 of [8], which asserts (in particular) the equivalence of two conditions labelled (1.59) and (1.60) there. The condition in (1.60) is implied by the definition of uniform rectifiability given in Definition 2.3 automatically, while (1.59) implies the WGL because of (1.73) on p. 27 of [8]. A more direct proof of the WGL for uniformly rectifiable sets can be given using Theorem 1.8 on p. 314 of [8], together with a mild (apparent) strengthening of uniform rectifiability which is equivalent to the formulation in Definition 2.3, and which can be derived from Section 17 of [6].

We should perhaps mention that many of the general results about uniform rectifiability in the literature (such as [6, 7, 8] were stated only in the context of *unbounded* sets. However, one can always reduce to that case by taking the union of a given bounded set with a d-plane that passes nearby. (This point came up before, in Chapter 7, just after the statement of Proposition 7.5.)

In the present situation, one could avoid some of the general theory of uniformly rectifiable sets by repeating the arguments from Chapters 8 and 9 to get big pieces of Lipschitz graphs starting from the big projections in Proposition 10.1. Specifically, one could use the big projections in Proposition 10.1 in place of the Lipschitz mapping h in Lemma 8.18 (in the context of the proof of Proposition 8.15). This would also avoid the need for the "lifting" $\widehat{S}$ of S in (8.22). However, the proof of

Proposition 10.1 that follows will use the fact that S satisfies the WGL (at least locally), and thereby relies on some of the general theory anyway.

Indeed, the task of proving Proposition 10.1 is made much easier by the knowledge that S^* is locally uniformly rectifiable. Let S and $B = B(x_0, r_0)$ be as in the statement of Proposition 10.1. (These assumptions will be in force for the rest of the chapter.) By the same argument as above, there is a uniformly rectifiable Ahlfors-regular set F which is compact and satisfies

$$(10.4) \qquad S^* \cap B\left(x_0, \frac{r_0}{3}\right) \subseteq F \subseteq S^* \cap B\left(x_0, \frac{2r_0}{3}\right).$$

LEMMA 10.5. *For each $\epsilon > 0$, there exist $x_1 \in F \cap B(x_0, \frac{r_0}{6})$, $r_1 \in [\frac{r_0}{C(\epsilon)}, \frac{r_0}{6}]$, and a d-plane P such that*

$$(10.6) \qquad \mathrm{dist}(x, P) \leq \epsilon r_1 \qquad \text{for all } x \in F \cap B(x_1, r_1).$$

The constant $C(\epsilon)$ depends only on ϵ, k, and n.

This is a direct consequence of the WGL, which F satisfies because it is uniformly rectifiable (as discussed above). Let us first recall the definition of the WGL. For each $\epsilon > 0$, let $\mathcal{B}(\epsilon)$ denote the set of "bad" pairs $(x_1, r_1) \in F \times (0, \mathrm{diam}\, F)$ for which there does not exist a d-plane P that satisfies (10.6). Also set

$$(10.7) \qquad \mathcal{B}(\epsilon, x, R) = \{(x_1, r_1) \in \mathcal{B}(\epsilon) : x_1 \in F \cap B(x, R) \text{ and } 0 < r_1 \leq R\}.$$

We say that the Ahlfors-regular set F satisfies the WGL (weak geometric lemma) if for each $\epsilon > 0$ there is a $C > 0$ such that

$$(10.8) \qquad \iint_{\mathcal{B}(\epsilon, x, R)} dH^d(x_1) \frac{dr_1}{r_1} \leq C R^d$$

for all $x \in F$ and $0 < R < \mathrm{diam}\, F$. See [8], especially p. 26–27, for more information and further references. To prove Lemma 10.5, we apply (10.8) with $x = x_0$ and $R = \frac{r_0}{6}$. We then choose $C(\epsilon)$ so large (depending also on the Ahlfors-regularity constant for F) that

$$(10.9) \qquad H^d\left(F \cap B\left(x_0, \frac{r_0}{6}\right)\right) \int_{C(\epsilon)^{-1} r_0}^{\frac{r_0}{6}} \frac{dr_1}{r_1} > C \left(\frac{r_0}{6}\right)^d,$$

where C is as in (10.8). This inequality ensures that $\mathcal{B}(\epsilon, x_0, \frac{r_0}{6})$ cannot contain all (x_1, r_1) with $x_1 \in F \cap B(x_0, r_0/6)$ and $\frac{r_0}{C(\epsilon)} \leq r_1 \leq r_0/6$, because of (10.8), and this is exactly what we want for Lemma 10.5.

From now on we let x_1 and r_1 be as in Lemma 10.5. To prove Proposition 10.1 it suffices to show the following.

LEMMA 10.10. *If $\epsilon > 0$ is small enough, depending only on k and n, then*

$$(10.11) \qquad \pi(F \cap B(x_1, r_1)) \supseteq P \cap B(x_1, \tfrac{r_1}{2})$$

when x_1, r_1, and P are as above in Lemma 10.5, and where π denotes the orthogonal projection onto P.

To prove this, set

$$B_1 = B(x_1, r_1),$$

10. BIG PROJECTIONS

and assume to the contrary that there is a $\xi \in P \cap (\frac{1}{2} B_1)$ such that $\pi^{-1}(\xi) \cap F \cap B_1 = \emptyset$. We want to show that if $\epsilon > 0$ is small enough, then we can construct a mapping ϕ whose existence contradicts the quasiminimality of S.

Let us first check that

(10.12) $$H^d\big(\{x \in F \cap B_1 : \tfrac{r_1}{2} \leq |\pi(x) - \pi(x_1)| \leq \tfrac{r_1}{2} + \epsilon r_1\}\big) \leq C \epsilon r_1^d.$$

This follows from (10.6) and the Ahlfors regularity of F. Specifically, (10.6) says that all of the elements of $F \cap B_1$ lie within ϵr_1 of P, and this implies that

$$\{x \in F \cap B_1 : \tfrac{r_1}{2} \leq |\pi(x) - \pi(x_1)| \leq \tfrac{r_1}{2} + \epsilon r_1\}$$

can be covered by $C \epsilon^{-(d-1)}$ balls of radius ϵr_1 (since P is a d-plane). The H^d measure of the intersection of F with any such ball is $\leq C(\epsilon r_1)^d$, by Ahlfors regularity, and (10.12) follows by summing over the balls.

Next, let ϕ_1 be a C-Lipschitz function on $\mathbf{R}^n$ with the following properties:

(10.13) $$\phi_1(x) = x \quad \text{when } x \in \mathbf{R}^n \setminus B_1, \text{ and also}$$
$$\text{when } |\pi(x) - \pi(x_1)| \geq \frac{r_1}{2} + \epsilon r_1;$$

(10.14) $$\phi_1(B_1) \subseteq B_1;$$

(10.15) $$\pi(\phi_1(x)) = \pi(x) \quad \text{for all } x \in \mathbf{R}^n;$$

and

(10.16) $$\phi_1(x) = \pi(x) \quad \text{for all } x \in T,$$

where

(10.17) $$T = \{x \in \mathbf{R}^n : \operatorname{dist}(x, P) \leq \epsilon r_1 \text{ and } |\pi(x) - \pi(x_1)| \leq \tfrac{r_1}{2}\}.$$

In other words, T is a slightly thickened version of the (flat) d-dimensional closed disk

$$\{u \in P : |u - \pi(x_1)| \leq \tfrac{r_1}{2}\},$$

and ϕ_1 projects T onto this flat disk without otherwise being too different from the identity mapping. It is not difficult to produce such a mapping ϕ_1, and with uniformly bounded Lipschitz constant. (Note that T lies well inside B_1, at least if ϵ is small enough.)

We are primarily interested in the restriction of ϕ_1 to $S^* \cap B_1$. Note that

(10.18) $$S^* \cap B_1 = F \cap B_1,$$

because of (10.4) and the inclusion

(10.19) $$B_1 \subseteq B(x_0, \tfrac{r_0}{3})$$

(which itself comes from the choice of x_1, r_1 in Lemma 10.5). If

$$x \in S^* \cap B_1 \quad \text{and} \quad |\pi(x) - \pi(x_1)| \geq \frac{r_1}{2} + \epsilon r_1,$$

then $\phi_1(x) = x$. We do not know exactly what ϕ_1 does when

$$x \in S^* \cap B_1 \quad \text{and} \quad \frac{r_1}{2} < |\pi(x) - \pi(x_1)| < \frac{r_1}{2} + \epsilon r_1,$$

but (10.12) guarantees that this set is pretty small, and so we shall not worry about it too much. If

$$x \in S^* \cap B_1 \quad \text{and} \quad |\pi(x) - \pi(x_1)| \leq \frac{r_1}{2},$$

then (10.6) says that $x \in T$, and $\phi_1(x) = \pi(x)$. This makes it easier for us to apply a second mapping ϕ_2 to send

$$\phi_1\big(\{x \in S^* \cap B_1 : |\pi(x) - \pi(x_1)| \leq \tfrac{r_1}{2}\}\big)$$

to a $(d-1)$-dimensional sphere (and hence a set of H^d-measure 0), which is our next task.

Let θ denote the mapping from $P \cap \overline{B}\big(\pi(x), \tfrac{r_1}{2}\big) \setminus \{\xi\}$ to $P \cap \partial B\big(\pi(x_1), \tfrac{r_1}{2}\big)$ which is the usual radial projection with center ξ. Define ϕ_2 first by setting it to be equal to the identity mapping on ∂T, and to be equal to θ on

(10.20) $$D = P \cap \overline{B}\big(\pi(x_1), \frac{r_1}{2}\big) \cap \pi(S^* \cap B_1).$$

More precisely, D does not contain ξ because of the way that ξ was chosen at the beginning of the proof (of Lemma 10.10), and so D is contained in the domain of θ. Also, D is a compact set, because S^* is relatively closed in U (the open set from Proposition 10.1 in which everything takes place), because B_1 is contained in U (since $B_1 \subseteq B(x_0, \tfrac{r_0}{3})$, as in (10.19)), and because of (10.6), which permits us to rewrite (10.20) as

$$D = P \cap \overline{B}\big(\pi(x_1), \frac{r_1}{2}\big) \cap \pi\big(S^* \cap \overline{B}(x_1, \tfrac{2r_1}{3})\big),$$

at least if ϵ is not too large. (In applying (10.6), we are also implicitly using (10.18).) The compactness of D implies that

$$\text{dist}(D, \xi) > 0,$$

so that θ is Lipschitz on D. We do not have any control on the Lipschitz constant, because we have no control on the distance from D to ξ.

From this we may conclude that our initial definition of ϕ_2 on $\partial T \cup D$ is Lipschitz, since the definitions on ∂T and D are compatible with each other at the places where they overlap (and because ∂T and $D \subseteq P$ intersect "transversally"). Next, we extend ϕ_2 to a Lipschitz mapping from T into itself. It is not hard to choose an extension which takes values in T, rather than just $\mathbf{R}^n$, and in any case this can be corrected afterwards using a Lipschitz retraction from T onto itself. Finally, we set $\phi_2(x) = x$ for $x \in \mathbf{R}^n \setminus T$, so that ϕ_2 becomes a Lipschitz mapping from $\mathbf{R}^n$ to itself (since ϕ_2 is equal to the identity on ∂T already, by the first step above).

Define $\phi : \mathbf{R}^n \to \mathbf{R}^n$ by $\phi = \phi_2 \circ \phi_1$. This is the Lipschitz mapping that we want to use to test (and then contradict) the quasiminimality of S.

Let $W = \{x \in \mathbf{R}^n : \phi(x) \neq x\}$, as usual. It is not difficult to check that

(10.21) $$W \cup \phi(W) \subseteq B_1;$$

this uses (10.13), (10.14), the facts that $\phi_2(T) \subseteq T$ and ϕ_2 is equal to the identity mapping on the complement of T (by construction), and the inclusion $T \subseteq B_1$. From (10.21) one can also verify that ϕ satisfies the usual requirements (1.5), (1.6). For this one should remember that $r_0 < \delta$ and $B = B(x_0, r_0) \subseteq U$, as in the statement of Proposition 10.1, while $B_1 = B(x_1, r_1) \subseteq \tfrac{1}{3} B$ and $r_1 \leq \tfrac{r_0}{6}$, by the conditions in Lemma 10.5. The homotopy condition (1.7) for ϕ can be derived as in Remark 1.10.

In short, ϕ meets the conditions required in the definition of a quasiminimizer (in Chapter 1). This permits us to apply (1.8) to obtain that

(10.22) $$H^d(S \cap W) \leq k H^d(\phi(S \cap W)).$$

We want to estimate both sides of this inequality.

The main point of the construction of ϕ is that

(10.23) $$H^d(\phi(S^* \cap T)) = 0$$

(and, in fact, $\phi(S^* \cap T)$ has finite H^{d-1}-measure). This property holds because

(10.24) $$\phi(S^* \cap T) = \phi_2(\phi_1(S^* \cap T))$$
$$\subseteq \phi_2(P \cap \overline{B}(\pi(x_1), \tfrac{r_1}{2}) \cap \pi(S^* \cap B_1))$$
$$\subseteq P \cap \partial B(\pi(x_1), \tfrac{r_1}{2}).$$

The first inclusion in (10.24) uses (10.16) and the definition (10.17) of T, while the second comes from the definition of ϕ_2 (i.e., the fact that $\phi_2 = \theta$ on D, where D is as in (10.20)).

It follows from (10.23) and the definition of W that H^d-almost every element of $S^* \cap T$ lies in W. Therefore

(10.25) $$H^d(S \cap W) \geq H^d(S^* \cap T) \geq H^d(S^* \cap B(x_1, \tfrac{r_1}{3})) \geq C^{-1} r_1^d.$$

The second inequality in (10.25) comes from the inclusion

$$S^* \cap T \supseteq S^* \cap B(x_1, \frac{r_1}{3}),$$

which itself follows from the definition (10.17) of T and the fact that x_1, r_1, and P satisfy (10.6). For this we also need ϵ to be reasonably small. The third inequality in (10.25) is a consequence of the local Ahlfors regularity of S^*, as in Proposition 4.1. (We can apply Proposition 4.1 here because $r_0 < \delta$ and $B = B(x_0, r_0) \subseteq U$, as in the statement of Proposition 10.1, and because of (10.19). One could also use the Ahlfors regularity of F instead.)

To get an upper bound for $H^d(\phi(S^* \cap W))$, let us first check that

(10.26) $$S^* \cap W \subseteq \{x \in S^* \cap B_1 : |\pi(x) - \pi(x_1)| \leq \tfrac{r_1}{2} + \epsilon r_1\}.$$

If x lies in $S^* \cap W$, then $\phi(x) \neq x$ in particular, so that either $\phi_1(x) \neq x$ or $\phi_1(x) = x$ and $\phi_2(x) \neq x$. In the first case x must lie in the right side of (10.26) by (10.13). In the second case the same is true because ϕ_2 is equal to the identity mapping on the complement of T, by construction (and because $S^* \cap T$ is automatically contained in right side of (10.26) by the definition (10.17) of T).

Using (10.18) we can rewrite (10.26) as

(10.27) $$S^* \cap W \subseteq \{x \in F \cap B_1 : |\pi(x) - \pi(x_1)| \leq \tfrac{r_1}{2} + \epsilon r_1\}.$$

We can convert this into

$$S^* \cap W \subseteq \{x \in F \cap B_1 : \tfrac{r_1}{2} \leq |\pi(x) - \pi(x_1)| \leq \tfrac{r_1}{2} + \epsilon r_1\} \cup (S^* \cap T),$$

because of (10.6). In other words, the elements x of $F \cap B_1$ that satisfy

$$|\pi(x) - \pi(x_1)| \leq \frac{r_1}{2}$$

automatically lie in T, by (10.6).

Since $H^d\big(\phi(S^* \cap T)\big) = 0$ (and $H^d(S \backslash S^*) = 0$, as (1.13)), we obtain that
$$H^d\big(\phi(S \cap W)\big) \leq H^d\big(\phi(\{x \in F \cap B_1 : \tfrac{r_1}{2} \leq |\pi(x) - \pi(x_1)| \leq \tfrac{r_1}{2} + \epsilon r_1\})\big).$$
On the other hand,
$$\phi(\{x \in F \cap B_1 : \tfrac{r_1}{2} \leq |\pi(x) - \pi(x_1)| \leq \tfrac{r_1}{2} + \epsilon r_1\})$$
$$= \phi_1\big(\{x \in F \cap B_1 : \tfrac{r_1}{2} \leq |\pi(x) - \pi(x_1)| \leq \tfrac{r_1}{2} + \epsilon r_1\}\big).$$
Indeed, if $|\pi(x) - \pi(x_1)| \geq \tfrac{r_1}{2}$, then $|\pi(\phi_1(x)) - \pi(x_1)| \geq \tfrac{r_1}{2}$, by (10.15), and hence $\phi_2\big(\phi_1(x)\big) = \phi_1(x)$, because ϕ_2 equals the identity mapping on the complement of T, by construction. This implies that
$$H^d\big(\phi(S \cap W)\big) \leq H^d\big(\phi_1(\{x \in F \cap B_1 : \tfrac{r_1}{2} \leq |\pi(x) - \pi(x_1)| \leq \tfrac{r_1}{2} + \epsilon r_1\})\big).$$
We can reduce this further to
$$H^d\big(\phi(S \cap W)\big) \leq C\, H^d\big(\{x \in F \cap B_1 : \tfrac{r_1}{2} \leq |\pi(x) - \pi(x_1)| \leq \tfrac{r_1}{2} + \epsilon r_1\}\big),$$
because ϕ_1 is Lipschitz with bounded constant. (We do not have a bound for the Lipschitz constant of ϕ_2, which is why we had to be careful to peel it off first.)

From here we conclude that

(10.28) $$H^d\big(\phi(S \cap W)\big) \leq C \epsilon\, r_1^d,$$

using (10.12). If we choose ϵ small enough, then (10.22), (10.25) and (10.28) cannot hold at the same time. This contradiction completes the proof of Lemma 10.10.

As we mentioned before, Proposition 10.1 follows from Lemma 10.10. The proof of Theorem 2.11 is now (finally) finished as well, because Proposition 10.1 was the only remaining ingredient for that, as explained at the beginning of the chapter.

CHAPTER 11

Restricted and Dyadic Quasiminimizers

In the previous chapters, we have obtained a lot of information about the structure of quasiminimizing sets, but we have said little about how quasiminimizing sets might arise. A basic point is that minimizers of functionals like

$$(11.1) \qquad J(S) = \int_S f(x)\, dH^d(x)$$

give rise to quasiminimizers in the sense of Chapter 1, at least if f is a positive (measurable) function that is bounded and bounded away from 0, and if the class of competitors allowed for the notion of "minimizer" is rich enough to accommodate the usual deformations of S (as in Definition 1.9).

Under what conditions do minimizers exist for a functional like $J(S)$ (and for a reasonably rich class of competitors)? As usual for this type of situation, in order to have the existence of minimizers, one should ask that f be lower semicontinuous, in addition to the conditions mentioned above. Let us also imagine that we have some kind of topological or boundary conditions for the set S, to prevent the minimization from degenerating to sets of measure 0. There remains the problem that Hausdorff measure does not behave very well under limits for sequences of sets, i.e., the value of $J(S)$ could "jump up" as a result of taking a limit. This is bad for the direct method in the calculus of variations, in which one tries to establish the existence of minimizers by taking limits of sequences of sets for which the value of J approaches the infimal value over all competitors.

This is a standard difficulty, and one which is often addressed through the use of currents, varifolds, or functions of bounded variation. Some basic references include [13], [14], and [27], for instance. Here we shall take a different tack, and consider "finite" (polyhedral) problems for which the existence of minima is immediate. It will then be important to have uniform bounds for the behavior of the minimizers, bounds that do not depend on the degree of resolution in the approximation.

In order to obtain such bounds, one can try to show that the minimizers for the finite problems are quasiminimizers in the sense of Definition 1.9, i.e., with respect to general deformations (and not just competitors in a fixed finite class), and with uniform bounds on the quasiminimizing constants k and δ from Definition 1.9. This would make the conclusions of Theorem 2.11 available, with uniform bounds as well.

We shall make use of this approach in Chapter 12. In the present chapter we discuss some tools for deriving bounds for quasiminimality constants when the set of competitors is initially restricted in some way. These tools will be applied to polyhedral minimizers for functionals like (11.1) in Sections 12.4 and 12.5. In Section 12.5 we shall also discuss the behavior of Hausdorff limits of sets when there are uniform bounds for Ahlfors regularity, uniform rectifiability, or the property of big pieces of Lipschitz graphs.

The next definition gives a notion of "restricted quasiminimizer" which will be useful for these purposes.

DEFINITION 11.2. Let $0 < d < n$, an open set $U \subseteq \mathbf{R}^n$, a set $A \subseteq U$, and numbers $k \in [1, +\infty)$ and $\delta \in (0, +\infty]$ be given. A (U, k, δ)-quasiminimizer restricted to A is a set $S \subseteq A$ that satisfies (1.1) and (1.2), and for which

(11.3) $$H^d(S \cap W) \leq k H^d(\phi(S \cap W))$$

holds whenever $\phi : \mathbf{R}^n \to \mathbf{R}^n$ is a Lipschitz mapping that satisfies (1.5), (1.6), (1.7) and

(11.4) $$\phi(S) \subseteq A.$$

(Here $W = \{x \in \mathbf{R}^n : \phi(x) \neq x\}$, as usual.)

The only difference between this definition and the original notion of quasiminimizer from Chapter 1 is that we have added the requirements that S be contained in A, and that the competitors $\phi(S)$ of S be contained in A as well. In particular, this definition would be equivalent to the previous one if A were all of $\mathbf{R}^n$, or all of U. A more significant situation would be to have A be a d-dimensional polyhedron, like the d-dimensional skeleton of some union of n-dimensional cubes in $\mathbf{R}^n$.

A natural variant of Definition 11.2 would be to add the requirement that for the mapping ϕ above, the homotopy implicit in the condition (1.7) also maps S into A for each "time" t. We have left this out for simplicity, but it would be somewhat more in the direction of "restricting to A". One could consider analogous adjustments throughout this chapter. (In practice, the difference frequently does not matter much by the end anyway.)

Proposition 11.13 below provides conditions under which quasiminimality restricted to a d-dimensional skeleton can be replaced by quasiminimality restricted to a larger set of full dimension. The next lemma concerns the situation where A is more like an n-dimensional set, and one wants to go from quasiminimality restricted to A to ordinary quasiminimality. In practice one might apply Proposition 11.13 first, to fill out a d-dimensional set A to an n-dimensional one, and then use Lemma 11.5 to get rid of the weaker restriction that remains.

In this chapter, U will always be an open subset of $\mathbf{R}^n$ (and d will be an integer, with $0 < d < n$).

LEMMA 11.5. Let $\delta > 0$, $C_1 \geq 1$, and $\delta_1 > 0$ be given, and let A be a subset of U. Assume that for each compact subset L of U with $L \cap A \neq \emptyset$ and $\operatorname{diam} L < \delta_1$ there is a Lipschitz mapping $h = h_L : \mathbf{R}^n \to \mathbf{R}^n$ with the following properties:

(11.6) $$h(x) = x \quad \text{for all } x \in A;$$

(11.7) $$h(x) \in A \quad \text{for all } x \in L;$$

(11.8) $$|h(x) - h(y)| \leq C_1 |x - y| \quad \text{for } x, y \in L;$$

and

(11.9) $$\operatorname{diam}(W_h \cup h(W_h)) < \frac{\delta}{2} \quad \text{and} \quad \operatorname{dist}(W_h \cup h(W_h), \mathbf{R}^n \setminus U) > 0,$$

where $W_h = \{x \in \mathbf{R}^n : h(x) \neq x\}$ (as usual). We also require that there be a continuous mapping $H : [0,1] \times \mathbf{R}^n \to \mathbf{R}^n$ which satisfies conditions analogous to those in (1.7) for this choice of h, i.e., $H(0,x) = x$ and $H(1,x) = h(x)$ for all $x \in \mathbf{R}^n$, $H(t,x)$ is Lipschitz in x for each $t \in [0,1]$, and if $\widehat{W}(h)$ denotes the *set of points in $\mathbf{R}^n$ of the form $H(t, x)$ or x, where $x \in \mathbf{R}^n$ and $t \in [0, 1]$ satisfy $H(t, x) \neq x$, then*

$$\operatorname{diam} \widehat{W}(h) < \frac{\delta}{2} \quad \text{and} \quad \operatorname{dist}(\widehat{W}(h), \mathbf{R}^n \backslash U) > 0.$$

Under these conditions, every (U, k, δ)-quasiminimizer restricted to A is an ordinary (unrestricted) $(U, \widetilde{k}, \widetilde{\delta})$-quasiminimizer, with $\widetilde{k} = C_1^d(1 + k)$ and $\widetilde{\delta} = \min(\delta_1, \frac{\delta}{2})$.

The existence of the retractions h may seem to be a complicated condition, but it is easy to check in some situations (like neighborhoods of unions of cubes of definite size), and it becomes simpler when $\delta = \delta_1 = +\infty$. See also Section 12.1.

To prove the lemma, let S be a (U, k, δ)-quasiminimizer restricted to A, and let $\phi : \mathbf{R}^n \to \mathbf{R}^n$ be a Lipschitz mapping that satisfies (1.5), (1.6), and (1.7), but with δ replaced by $\widetilde{\delta}$. We want to verify the quasiminimality condition (1.8) for ϕ and S, and with k replaced by $\widetilde{k}$. Set

$$W = \{x \in \mathbf{R}^n : \phi(x) \neq x\}.$$

We may as well assume that $W \cap S \neq \emptyset$, since otherwise (1.8) is trivial.

Put $F = W \cup \phi(W)$, and let L be the closure of F. Thus L intersects S, since F does. This implies that L intersects A, because $S \subseteq A$, since S is a quasiminimizer restricted to A. The diameter of F is strictly less than $\widetilde{\delta} \leq \delta_1$, because of (1.5) (with δ replaced by $\widetilde{\delta}$) and the definition of $\widetilde{\delta}$ in Lemma 11.5. Thus L has diameter strictly less than δ_1, and L is a compact subset of U, because of (1.6). The hypotheses of Lemma 11.5 provide us with a Lipschitz mapping $h = h_L : \mathbf{R}^n \to \mathbf{R}^n$ that satisfies (11.6)–(11.9). Put

$$\widetilde{\phi} = h \circ \phi.$$

We want to apply the restricted quasiminimality property of S to $\widetilde{\phi}$, and then use the result to derive the ordinary quasiminimality property for S and ϕ as in (1.8). To do this, we begin by verifying that $\widetilde{\phi}$ satisfies the requirements of Definition 11.2 (with respect to δ).

Clearly $\widetilde{\phi}$ is Lipschitz since ϕ and h are. Also, $\widetilde{\phi}(S) \subseteq A$. This comes from the way that we chose h. Specifically, if p is an arbitrary point in S, then either $\phi(p) = p$, or $\phi(p) \neq p$. If $\phi(p) = p$, then $\phi(p) \in A$ (since $S \subseteq A$, by assumption — see Definition 11.2), and $h(\phi(p)) = \phi(p) \in A$ by (11.6). If $\phi(p) \neq p$, then $p \in W$, $\phi(p) \in \phi(W) \subseteq F \subseteq L$, and $h(\phi(p)) \in A$, by (11.7). Thus $\widetilde{\phi}(p) = h(\phi(p))$ lies in A for all $p \in S$, and so (11.4) holds (with ϕ replaced with $\widetilde{\phi}$).

It remains to check that $\widetilde{\phi}$ satisfies the analogues of (1.5), (1.6), and (1.7). Set $\widetilde{W} = \{x \in \mathbf{R}^n : \widetilde{\phi}(x) \neq x\}$. Thus

$$\widetilde{W} \subseteq W \cup (W_h \backslash W),$$

as one can easily check. (If $\widetilde{\phi}(x) \neq x$, then either $\phi(x) \neq x$, or $\phi(x) = x$ and $h(x) \neq x$, and this is what the inclusion for $\widetilde{W}$ says.) This yields

$$\widetilde{\phi}(\widetilde{W}) \subseteq \widetilde{\phi}(W) \cup h(W_h) = h(\phi(W)) \cup h(W_h),$$

since ϕ is equal to the identity on $W_h \backslash W$. We can refine this a bit by observing that
$$h(\phi(W)) = h(\phi(W)\backslash W_h) \cup h(\phi(W) \cap W_h)$$
$$\subseteq \phi(W) \cup h(W_h)$$
(since h is equal to the identity on $\phi(W)\backslash W_h$). Combining this with the previous step we obtain that $\widetilde{\phi}(\widetilde{W}) \subseteq \phi(W) \cup h(W_h)$, and hence that
$$\widetilde{W} \cup \widetilde{\phi}(\widetilde{W}) \subseteq (W \cup \phi(W)) \cup (W_h \cup h(W_h)),$$
because of the earlier inclusion for $\widetilde{W}$. This shows that $\widetilde{W} \cup \widetilde{\phi}(\widetilde{W})$ lies at positive distance from $\mathbf{R}^n \backslash U$, because of (11.9) and (1.6). In other words, $\widetilde{\phi}$ satisfies (1.6).

Next we want to estimate the diameter of $\widetilde{W} \cup \widetilde{\phi}(\widetilde{W})$. By assumption, ϕ satisfies (1.5) with δ replaced by $\widetilde{\delta}$, so that $\operatorname{diam}(W \cup \phi(W)) < \widetilde{\delta}$. Also,
$$\operatorname{diam}(W_h \cup h(W_h)) < \frac{\delta}{2},$$
by (11.9). If $W \cup \phi(W)$ intersects $W_h \cup h(W_h)$, then we may conclude that
$$(11.10) \quad \operatorname{diam}(\widetilde{W} \cup \widetilde{\phi}(\widetilde{W})) \leq \operatorname{diam}((W \cup \phi(W)) \cup (W_h \cup h(W_h)))$$
$$\leq \operatorname{diam}(W \cup \phi(W)) + \operatorname{diam}(W_h \cup h(W_h))$$
$$< \widetilde{\delta} + \delta/2 \leq \delta,$$
since $\widetilde{\delta} \leq \delta/2$, by definition (as in Lemma 11.5). This would give (1.5) for $\widetilde{\phi}$ (and δ).

In fact, we may as well assume that $W \cup \phi(W)$ intersects $W_h \cup h(W_h)$, for the following reasons. If $\phi(S) \subseteq A$ already, then the quasiminimality inequality that we are after comes directly from (11.3). That is, we can apply Definition 11.2 to ϕ itself, without worrying about $\widetilde{\phi}$ at all, and there is nothing really to do. Thus we assume instead that $\phi(S)$ is not (wholly) contained in A. Because $S \subseteq A$ by assumption (see Definition 11.2), $\phi(S) \not\subseteq A$ implies that $\phi(W)$ is not a subset of A. Let p be any element of $\phi(W) \backslash A$. Then $p \in L$, since $\phi(W) \subseteq L$ by the definition of L (near the beginning of the proof). Thus $h(p) \in A$, by (11.7). This means that $h(p) \neq p$, since $p \notin A$, and hence that $p \in W_h$. Thus p lies in both $\phi(W)$ and W_h, and so that $W \cup \phi(W)$ intersects $W_h \cup h(W_h)$. This gives (11.10), and therefore (1.5) for $\widetilde{\phi}$ as well.

It remains to show that the analogue of (1.7) holds for $\widetilde{\phi}$ (and δ). That is, we would like to have a continuous one-parameter family of mappings $\widetilde{\phi}_t$, $0 \leq t \leq 1$, such that $\widetilde{\phi}_0$ is the identity, $\widetilde{\phi}_1 = \widetilde{\phi}$, each $\widetilde{\phi}_t$ is Lipschitz, and so that the analogue of (1.7c) holds (for δ). By assumption, ϕ itself admits such a family, with δ replaced by $\widetilde{\delta}$ in (1.7c), and we have a similar family for h, explicitly included in the hypotheses of Lemma 11.5 (and with δ in (1.7c) replaced with $\delta/2$). Set $\widetilde{\phi}_t = h_t \circ \phi_t$, where $h_t(x) = H(t,x)$, and $H(t,x)$ is as in the statement of Lemma 11.5. Thus $\widetilde{\phi}_0$ is the identity mapping, $\widetilde{\phi}_1 = h \circ \phi = \widetilde{\phi}$, and each $\widetilde{\phi}_t$ is Lipschitz, because of the corresponding properties for ϕ_t and h_t. As for (1.7c), let $\widehat{W}(\widetilde{\phi})$ denote the set of points in $\mathbf{R}^n$ which are of the form $\widetilde{\phi}_t(x)$ or x, where $x \in \mathbf{R}^n$ and $t \in [0,1]$ satisfy $\widetilde{\phi}_t(x) \neq x$. Define $\widehat{W}(\phi)$ and $\widehat{W}(h)$ similarly, using ϕ and h instead of $\widetilde{\phi}$.

Thus diam $\widehat{W}(\phi) < \widetilde{\delta}$ by our initial assumptions on ϕ, and diam $\widehat{W}(h) < \delta/2$, as in Lemma 11.5. On the other hand,

$$\widehat{W}(\widetilde{\phi}) \subseteq \widehat{W}(\phi) \cup \widehat{W}(h),$$

for exactly the same reason that $\widetilde{W} \cup \widetilde{\phi}(\widetilde{W})$ is contained in the union of $W \cup \phi(W)$ and $W_h \cup h(W_h)$, as before. (More precisely, for each $t \in [0, 1]$, one can treat $\widetilde{\phi}_t = h_t \circ \phi_t$ in exactly the same manner as $\widetilde{\phi} = h \circ \phi$ was treated before.) We also have that $\widehat{W}(\phi)$ and $\widehat{W}(h)$ intersect; this is because $\widehat{W}(\phi) \supseteq W \cup \phi(W)$ and $\widehat{W}(h) \supseteq W_h \cup h(W_h)$, by construction (take $t = 1$), and because we can assume that $W \cup \phi(W)$ and $W_h \cup h(W_h)$ intersect, as discussed above. Thus we are in the same situation as in (11.10), and we obtain that

$$\begin{aligned}\operatorname{diam}(\widehat{W}(\widetilde{\phi})) &\leq \operatorname{diam}(\widehat{W}(\phi) \cup \widehat{W}(h)) \\ &\leq \operatorname{diam}\widehat{W}(\phi) + \operatorname{diam}\widehat{W}(h) \\ &< \widetilde{\delta} + \tfrac{\delta}{2} \leq \delta.\end{aligned}$$

This is precisely what we need for (1.7c). Thus $\widetilde{\phi}$ satisfies (1.7) (with respect to δ), as desired.

To summarize, we have shown that $\widetilde{\phi} : \mathbf{R}^n \to \mathbf{R}^n$ is admissible for Definition 11.2. This permits us to apply (11.3) to obtain that

(11.11) $$\begin{aligned}H^d(S \cap \widetilde{W}) &\leq k H^d(\widetilde{\phi}(S \cap \widetilde{W})) \\ &\leq k\, C_1^d H^d(\phi(S \cap \widetilde{W})).\end{aligned}$$

For the second inequality in (11.11) we are using (11.8) to say that h is C_1-Lipschitz on $\phi(S \cap \widetilde{W})$. More precisely, in order to employ (11.8) in this manner we should verify that

$$\phi(S \cap \widetilde{W}) \subseteq L.$$

Let us first check that

$$S \cap \widetilde{W} \subseteq W.$$

If x lies in $S \backslash W$, then $\phi(x) = x$, by definition of W, and

$$\widetilde{\phi}(x) = h(\phi(x)) = h(x) = x.$$

The last equality follows from (11.6) and the fact that $x \in S \subseteq A$ (with $S \subseteq A$ because of Definition 11.2). Thus $x \in S \backslash W$ implies that $\widetilde{\phi}(x) = x$, and hence that $x \notin \widetilde{W}$. This proves that $S \cap \widetilde{W} \subseteq W$. From here we obtain that

$$\phi(S \cap \widetilde{W}) \subseteq \phi(W) \subseteq L,$$

by the definition of L (given at the beginning of the proof of Lemma 11.5). Thus (11.8) does indeed imply that h is C_1-Lipschitz on $\phi(S \cap \widetilde{W})$, and the second inequality in (11.11) follows (using also (0.4)).

Our eventual goal is to show that ϕ satisfies the quasiminimality condition (1.8) (with $\widetilde{k}$ instead of $\widetilde{k}$). To this end we wish to have an estimate like (11.11), but with $\widetilde{W}$ replaced with W.

Set $S_1 = (S \cap W) \setminus \widetilde{W}$. If $x \in S_1$, then $\phi(x) \neq x$, but $\widetilde{\phi}(x) = x$ (by definition of W and $\widetilde{W}$). Since $\widetilde{\phi}(x) = h(\phi(x))$, we conclude that $S_1 = h(\phi(S_1))$. We also have that $\phi(S_1) \subseteq \phi(W) \subseteq L$, as above, so that h is C_1-Lipschitz on $\phi(S_1)$. Thus

$$(11.12) \qquad H^d(S_1) = H^d(h(\phi(S_1))) \leq C_1^d H^d(\phi(S_1))$$
$$\leq C_1^d H^d(\phi(S \cap W)).$$

From (11.11) we have that

$$H^d(S \cap \widetilde{W}) \leq k\, C_1^d H^d(\phi(S \cap W)),$$

since $S \cap \widetilde{W} \subseteq W$, as observed above. Combining this with (11.12) we get that

$$H^d(S \cap W) \leq H^d(S \cap \widetilde{W}) + H^d(S_1)$$
$$\leq C_1^d (k+1)\, H^d(\phi(S \cap W)).$$

This is the same as (1.8), but with k in (1.8) replaced with $\widetilde{k} = C_1^d(k+1)$. Thus S is a $(U, \widetilde{k}, \delta)$-quasiminimizer, and the proof of Lemma 11.5 is complete.

Next we want to consider "dyadic restricted quasiminimizers", which are restricted quasiminimizers in which the set A is taken to be a part of a d-dimensional dyadic "skeleton". To make this precise, fix an integer j, and let Δ_j denote the set of dyadic cubes in $\mathbf{R}^n$ of sidelength 2^{-j}. Thus Δ_j consists of the cubes in $\mathbf{R}^n$ which are obtained from $[0, 2^{-j}]^n$ by translations in $(2^{-j}\mathbf{Z})^n$. (In this discussion, "cubes" will always be closed cubes.) For each integer $0 \leq m \leq n$, let $\mathcal{S}_{j,m}$ denote the union of all the m-dimensional faces of cubes $Q \in \Delta_j$. This is the "m-dimensional dyadic skeleton" of $\mathbf{R}^n$. Note that $\mathcal{S}_{j,n} = \mathbf{R}^n$ and $\mathcal{S}_{j,0} = \mathbf{Z}^n$. We shall mostly be concerned with intermediate dimensions, however.

The next result provides conditions under which one can pass from a set being a restricted quasiminimizer relative to a subset of a d-dimensional dyadic skeleton, to being a restricted quasiminimizer relative to a larger set of full dimension.

PROPOSITION 11.13. *Fix $j \in \mathbf{Z}$, and suppose that $A \subseteq U$ is a finite union of dyadic cubes in Δ_j, and that $\mathrm{dist}(A, \mathbf{R}^n \setminus U) > 2\sqrt{n}\, 2^{-j}$. If S is a (U, k, δ) quasiminimizer for H^d restricted to $A \cap \mathcal{S}_{j,d}$, and if $\widetilde{\delta} = \delta - 4\sqrt{n}\, 2^{-j} > 0$, then S is a $(U, Ck, \widetilde{\delta})$-quasiminimizer for H^d restricted to A. The constant $C \geq 1$ depends only on n and d.*

As usual, U is an open subset of $\mathbf{R}^n$. It is very important here that the constant C does not depend on j, i.e., so that the estimates that do not depend on the degree of "resolution" (or scale of polyhedral approximation). The proof of Proposition 11.13 will rely on the following variant of Proposition 3.1.

LEMMA 11.14. *Fix $j \in \mathbf{Z}$, and let E be a compact subset of $\mathbf{R}^n$ such that $H^d(E) < +\infty$. Denote by $\mathcal{N}_j(E)$ the union of all the cubes $Q \in \Delta_j$ that touch a cube in Δ_j which intersects E. Then there is a Lipschitz mapping $f : \mathbf{R}^n \to \mathbf{R}^n$ with the following properties:*

(11.15) $\qquad\qquad f(x) = x \quad \text{for} \ \ x \in \mathbf{R}^n \setminus \mathcal{N}_j(E),$

(11.16) $\qquad\qquad f(x) = x \quad \text{for all} \ \ x \in \mathcal{S}_{j,d},$

(11.17) $\qquad\qquad f(E) \subseteq \mathcal{S}_{j,d},$

(11.18) $\qquad\qquad f(Q) \subseteq Q \quad \text{for all} \ \ Q \in \Delta_j,$

and

(11.19) $$H^d\big(f((E\cap Q)\backslash \mathcal{S}_{j,d})\big) \leq CH^d((E\cap Q)\backslash \mathcal{S}_{j,d}) \quad \text{for all} \quad Q \in \Delta_j.$$

Again, this constant C depends only on n and d, and not on j or E.

Lemma 11.14 is very similar to Proposition 3.1. We shall not repeat the construction, but only highlight the key differences.

In Proposition 3.1 we worked relative to a fixed cube Q containing E, while now there is no such distinguished cube. In particular, we do not have a boundary term like the ∂Q in (3.4), and there is nothing like the special treatment of cubes that meet the boundary as there was before. However, analogous to (3.2) we now have the condition (11.15). We want to verify now that the construction in Chapter 3 gives (11.15) automatically, at least if one never chooses the various extensions (after (3.21)) too foolishly.

More precisely, if, in the context of (3.21), the intersection of $\phi_m(E)$ with $T\backslash \partial T$ is empty, then one should not worry about the radial projections $\theta_{\xi,T}$ on T, but instead simply take ψ_{m-1} to be equal to the identity mapping on T. Similarly, for the subsequent extensions to higher-dimensional cubes (shortly after (3.21) in Chapter 3), one should use the identity mapping whenever possible, i.e., when it was used already for all of the boundary faces of the (higher-dimensional) cube in question. These conventions will be in force for the rest of the discussion of the proof of Lemma 11.14.

To show that (11.15) holds under these conditions, let us mention first that if x is any point in $\mathbf{R}^n$ and Q is a cube in Δ_j which contains x, then the construction in Chapter 3 gives $\phi_m(x) \in Q$ for each m, $d \leq m \leq n$, where ϕ_m is as discussed between the statements of Proposition 3.1 and Lemma 3.10. This is not hard to see, and indeed it comes (via induction) from (3.9) and (3.13). (Compare also with (3.5).)

Fix m for a moment, and suppose that T is an m-dimensional cube, a face of some cube in Δ_j, with the property that $\phi_m(E)$ intersects the interior of T. In this situation, T must be a face of a cube $Q \in \Delta_j$ which intersects E. That is, if $x \in E$ has the property that $\phi_m(x)$ lies in the interior of T, and if $Q \in \Delta_j$ contains x, then $\phi_m(x) \in Q$, as in the preceding paragraph, and so $T \subseteq Q$ (since $\phi_m(x)$ lies in the interior of T).

Given this, let us check that if $\phi_m(y) \neq y$ for some m between d and n, then y must lie in a cube $Q \in \Delta_j$ which touches another cube in Δ_j that intersects E. We use induction on m, with the $m=n$ case being trivial (since ϕ_n is the identity by construction, as mentioned just above (3.9)). If $\phi_{m-1}(y) \neq y$, then either $\phi_m(y) \neq y$, in which case we are safe, by induction, or $\phi_m(y) = y$ and $\psi_{m-1}(y) \neq y$. (Recall that $\phi_{m-1} = \psi_{m-1} \circ \phi_m$, as in (3.9).) Choose $Q \in \Delta_j$ so that $y \in Q$. Assuming $\psi_{m-1}(y) \neq y$, there must be an m-dimensional face T of Q such that $\phi_m(E)$ intersects the interior of T; otherwise we could take ψ_{m-1} to be equal to the identity mapping on Q, and we would be required to do this, by the conventions discussed previously. This would contradict $\psi_{m-1}(y) \neq y$. Thus Q must contain such an m-dimensional face T, and T is contained in a cube $Q' \in \Delta_j$ which intersects E, by the observation in the preceding paragraph. To summarize, y is contained in Q, $Q \cap Q' \neq \emptyset$, and $Q' \cap E \neq \emptyset$, and this is exactly what we need for (11.15).

There is another minor difference between Lemma 11.14 and Proposition 3.1, which is that we look at the intersection of E with $Q\backslash \mathcal{S}_{j,d}$ in (11.19), rather than the intersection of E with Q, as in (3.6). While it is not true that (11.19) follows directly from (3.6), one can obtain it from exactly the same kind of argument as in Chapter 3. Keep in mind that the elements of $\mathcal{S}_{j,d}$ are held fixed throughout the construction, as in (11.16) and (3.3) (and thus do not really contribute to (3.6) in a tricky way). The main point in deriving the estimate (3.6) was the averaging inequality in Lemma 3.22, for which the particular choice of a set F was not important, so that very similar arguments can be used to obtain (11.19).

The remaining conditions (11.16)-(11.18) can all be treated in practically the same manner as in the proof of Proposition 3.1. This completes our discussion of the proof of Lemma 11.14.

Let us return now to Proposition 11.13. Let S be a quasiminimizer restricted to $A \cap \mathcal{S}_{j,d}$ (using δ). We want to show that S is also a quasiminimizer restricted to A (with respect to $\widetilde{\delta}$). To this end we let a Lipschitz mapping $\phi : \mathbf{R}^n \to \mathbf{R}^n$ be given, and we assume that ϕ satisfies (1.5)–(1.7) with $\widetilde{\delta} = \delta - 4\sqrt{n}2^{-j}$ instead of δ, and also

$$(11.20) \qquad \phi(S) \subseteq A.$$

Set $W = \{x \in \mathbf{R}^n : \phi(x) \neq x\}$ and

$$E = \overline{\phi(S \cap W)}.$$

Let f be as in Lemma 11.14, with this choice of E, and with j as in Proposition 11.13. We assume that

$$S \cap W \neq \emptyset,$$

since otherwise there is nothing to do, i.e., the quasiminimizing inequality (11.3) is trivial. Our first task is to show that $\widetilde{\phi} = f \circ \phi$ is an acceptable deformation for S as a quasiminimizer restricted to $A \cap \mathcal{S}_{j,d}$. We shall then use the analogue of (1.8) for $\widetilde{\phi}$ to derive a similar inequality for ϕ.

Put $\widetilde{W} = \{x \in \mathbf{R}^n : \widetilde{\phi}(x) \neq x\}$. Let us check that

$$\widetilde{W} \cup \widetilde{\phi}(\widetilde{W}) \subseteq W \cup \phi(W) \cup \mathcal{N}_j(E).$$

If $x \in \widetilde{W}$, so that $\widetilde{\phi}(x) \neq x$, then either $\phi(x) \neq x$, in which case $x \in W$ by definition, or $\phi(x) = x$ and $f(x) \neq x$, so that $x \in \mathcal{N}_j(E)$ because of (11.15). This shows that $\widetilde{W} \subseteq W \cup \mathcal{N}_j(E)$. Similarly, if $\widetilde{\phi}(x) \neq x$, then either $f(\phi(x)) \neq \phi(x)$, in which case $\phi(x) \in \mathcal{N}_j(E)$ (by (11.15)) and $\widetilde{\phi}(x) = f(\phi(x)) \in \mathcal{N}_j(E)$ (by (11.18) and the definition of $\mathcal{N}_j(E)$), or $f(\phi(x)) = \phi(x)$ and $\phi(x) \neq x$, in which event $x \in W$ and $\widetilde{\phi}(x) = \phi(x) \in \phi(W)$. This shows that $\widetilde{\phi}(\widetilde{W}) \subseteq \phi(W) \cup \mathcal{N}_j(E)$, and the inclusion for $\widetilde{W} \cup \widetilde{\phi}(\widetilde{W})$ follows.

Every element of $\mathcal{N}_j(E)$ lies at distance $\leq 2\sqrt{n}2^{-j}$ from $\phi(S \cap W)$, by the definition of $\mathcal{N}_j(E)$ (from Lemma 11.14), and because $E = \overline{\phi(S \cap W)}$. Hence every element of $\mathcal{N}_j(E)$ lies at distance $\leq 2\sqrt{n}2^{-j}$ from $W \cup \phi(W)$, since

$$\phi(S \cap W) \subseteq W \cup \phi(W)$$

trivially. (If $\phi(S \cap W)$ were empty (which it is not, by assumption), then E and $\mathcal{N}_j(E)$ would be too, and there would be nothing to do anyway.) This implies that

$$\operatorname{diam}(\widetilde{W} \cup \widetilde{\phi}(\widetilde{W})) \leq \operatorname{diam}(W \cup \phi(W)) + 4\sqrt{n}2^{-j}.$$

We also have that $\operatorname{diam}(W \cup \phi(W)) < \widetilde{\delta}$, because of the requirement that ϕ satisfy (1.5) with δ replaced by $\widetilde{\delta}$. Therefore

$$\operatorname{diam}(\widetilde{W} \cup \widetilde{\phi}(\widetilde{W})) < \delta,$$

because of the definition of $\widetilde{\delta}$ as $\delta - 4\sqrt{n}2^{-j}$ in Proposition 11.13. Thus $\widetilde{\phi}$ satisfies (1.5).

We also have that $\widetilde{\phi}$ satisfies (1.6), i.e., that

$$\operatorname{dist}(\widetilde{W} \cup \widetilde{\phi}(\widetilde{W}), \mathbf{R}^n \backslash U) > 0.$$

To see this, it is enough to know that

$$\operatorname{dist}(\mathcal{N}_j(E), \mathbf{R}^n \backslash U) > 0,$$

since

$$\operatorname{dist}(W \cup \phi(W), \mathbf{R}^n \backslash U) > 0$$

holds by the requirement (1.6) for ϕ. On the other hand, we assumed in Proposition 11.13 that

$$\operatorname{dist}(A, \mathbf{R}^n \backslash U) > 2\sqrt{n}2^{-j},$$

while every element of $\mathcal{N}_j(E)$ lies at distance $\leq 2\sqrt{n}2^{-j}$ from $\phi(S \cap W)$, as mentioned above. Every element of $\mathcal{N}_j(E)$ therefore lies at distance $\leq 2\sqrt{n}2^{-j}$ from A, because of (11.20). This ensures that the distance from $\mathcal{N}_j(E)$ to $\mathbf{R}^n \backslash U$ is positive, as desired, because of the assumed lower bound for $\operatorname{dist}(A, \mathbf{R}^n \backslash U)$. This proves that $\widetilde{\phi}$ satisfies (1.6).

Next, we want to check that $\widetilde{\phi}$ satisfies (1.7). Since ϕ satisfies (1.7) (with δ replaced by $\widetilde{\delta}$), there is a continuous one-parameter family of mappings ϕ_t, $0 \leq t \leq 1$, such that ϕ_0 is the identity mapping, $\phi_1 = \phi$, each ϕ_t is Lipschitz, and

$$\operatorname{diam} \widehat{W}(\phi) < \widetilde{\delta} \quad \text{and} \quad \operatorname{dist}(\widehat{W}(\phi), \mathbf{R}^n \backslash U) > 0.$$

Here $\widehat{W}(\phi)$ is, as in (1.7), the set of points in $\mathbf{R}^n$ which are of the form x or $\phi_t(x)$, where $x \in \mathbf{R}^n$ and $t \in [0,1]$ satisfy $\phi_t(x) \neq x$. Define $f_t : \mathbf{R}^n \to \mathbf{R}^n$, $0 \leq t \leq 1$, by

$$f_t(x) = (1-t)x + tf(x).$$

Thus f_t is a continuous one-parameter family of mappings such that f_0 is the identity mapping, $f_1 = f$, each f_t is Lipschitz, and each f_t satisfies (11.15) and (11.18), $0 \leq t \leq 1$. (For (11.18) this uses the fact that cubes are convex.) Set

$$\widetilde{\phi}_t = f_t \circ \phi_t, \quad 0 \leq t \leq 1.$$

From the properties of ϕ_t and f_t we have that $\widetilde{\phi}_0$ is the identity mapping, $\widetilde{\phi}_1 = \widetilde{\phi}$, and each $\widetilde{\phi}_t$ is Lipschitz. Let $\widehat{W}(\widetilde{\phi})$ be the set of points in $\mathbf{R}^n$ of the form x or $\widetilde{\phi}_t(x)$, where $x \in \mathbf{R}^n$ and $t \in [0,1]$ satisfy $\widetilde{\phi}_t(x) \neq x$. It is not hard to check that

$$\widehat{W}(\widetilde{\phi}) \subseteq \widehat{W}(\phi) \cup \mathcal{N}_j(E),$$

for the same reasons that $\widetilde{W} \cup \widetilde{\phi}(\widetilde{W})$ is contained in $W \cup \phi(W) \cup \mathcal{N}_j(E)$ as before. More precisely, for each fixed t one can apply the same arguments to $\widetilde{\phi}_t$, f_t, and ϕ_t as were used previously for $\widetilde{\phi}$, f, and ϕ, and then take the union over $t \in [0,1]$. For these arguments to work one needs to know that each f_t, $0 \leq t \leq 1$, satisfies the analogues of (11.15) and (11.18), and also maps $\mathcal{N}_j(E)$ to itself. We already mentioned that f_t inherits the analogues of (11.15) and (11.18) from the original versions for f itself, while

$$f_t(\mathcal{N}_j(E)) \subseteq \mathcal{N}_j(E), \quad t \in [0,1],$$

follows from (11.18) for f_t and the fact that $\mathcal{N}_j(E)$ is a union of elements of Δ_j, by definition. (See Lemma 11.14.)

Once we know that $\widehat{W}(\widetilde{\phi})$ is contained in $\widehat{W}(\phi) \cup \mathcal{N}_j(E)$, we are able to conclude that

$$\operatorname{diam} \widehat{W}(\widetilde{\phi}) < \delta \quad \text{and} \quad \operatorname{dist}\left(\widehat{W}(\widetilde{\phi}), \mathbf{R}^n \backslash U\right) > 0,$$

as required in (1.7c). These inequalities can be derived from the corresponding statements for ϕ (with δ replaced by $\widetilde{\delta}$) in practically the same manner that (1.5) and (1.6) for $\widetilde{\phi}$ were derived from their counterparts for ϕ before. (For the bound on $\operatorname{diam} \widehat{W}(\widetilde{\phi})$, note that $\phi(S \cap W) \subseteq \widehat{W}(\phi)$, because $\phi(S \cap W) \subseteq W \cup \phi(W)$, as before, while $W \cup \phi(W) \subseteq \widehat{W}(\phi)$ holds by the definition of $\widehat{W}(\phi)$ (simply by taking $t = 1$). This permits one to say that every element of $\mathcal{N}_j(E)$ lies at distance $\leq 2\sqrt{n}2^{-j}$ from $\widehat{W}(\phi)$, and then the rest of the argument is the same as it was previously. For the statement that the distance from $\widehat{W}(\widetilde{\phi})$ to the complement of U is positive, essentially no changes in the earlier argument are needed at all.) Thus we have that $\widetilde{\phi}$ satisfies (1.7).

Of course $\widetilde{\phi} = f \circ \phi$ is automatically Lipschitz, and so the last point about $\widetilde{\phi}$ (for showing that it is an acceptable deformation for S as a quasiminimizer restricted to $A \cap \mathcal{S}_{j,d}$) is that

$$\widetilde{\phi}(S) \subseteq A \cap \mathcal{S}_{j,d}.$$

To verify this, notice first that S itself must be contained in $A \cap \mathcal{S}_{j,d}$, since S is assumed to be a quasiminimizer restricted to $A \cap \mathcal{S}_{j,d}$. (See the statement of Proposition 11.13 and Definition 11.2.) Let x be any element of S. If $\phi(x) = x$, then $\widetilde{\phi}(x) = f(\phi(x)) = f(x) = x$, because of (11.16) and the fact that $x \in S \subseteq \mathcal{S}_{j,d}$. In particular, $\widetilde{\phi}(x)$ lies in $A \cap \mathcal{S}_{j,d}$ in this case (as desired), since x does (because $x \in S$). If $\phi(x) \neq x$, then $x \in S \cap W$. Thus $\phi(x) \in E$ (by the definition of E as the closure of $\phi(S \cap W)$), and $\widetilde{\phi}(x) = f(\phi(x)) \in \mathcal{S}_{j,d}$, by (11.17). We also have that $\phi(x) \in A$, because of (11.20). This implies that $\widetilde{\phi}(x) = f(\phi(x))$ lies in A too, because of (11.18) and the assumption in Proposition 11.13 that A be a union of cubes in Δ_j. Thus we have that $\widetilde{\phi}(x) \in A \cap \mathcal{S}_{j,d}$ in this second case as well. This proves that $\widetilde{\phi}(x)$ lies in $A \cap \mathcal{S}_{j,d}$ whenever $x \in S$, so that $\widetilde{\phi}(S) \subseteq A \cap \mathcal{S}_{j,d}$.

In conclusion, we obtain that $\widetilde{\phi}$ is an acceptable deformation for the (U, k, δ)-quasiminimality property for S restricted to $A \cap \mathcal{S}_{j,d}$. This enables us to apply (11.3) to obtain that

(11.21) $$H^d(S \cap \widetilde{W}) \leq k H^d\left(\widetilde{\phi}(S \cap \widetilde{W})\right).$$

We want to use this to establish an analogous inequality for ϕ instead of $\widetilde{\phi}$.

Set

$$S_0 = \{x \in S \cap W : \phi(x) \in \mathbf{R}^n \backslash \mathcal{S}_{j,d}\}.$$

Thus

$$\phi(S_0) = \phi(S \cap W) \backslash \mathcal{S}_{j,d}$$

automatically. Let us check that
$$\phi(S_0) = E \backslash \mathcal{S}_{j,d}.$$
Because E was defined to be the closure of $\phi(S \cap W)$, the only issue is to show that
$$E \backslash \big(\phi(S \cap W) \cup \mathcal{S}_{j,d}\big) = \emptyset.$$
Remember that
$$\text{dist}(W, \mathbf{R}^n \backslash U) > 0,$$
by (1.6) (for ϕ), and that S is relatively closed in U, because of (1.1). (More precisely, (1.1) was included in the requirements of Definition 11.2, and hence is implicitly part of the hypotheses of Proposition 11.13.) From here it follows $S \cap \overline{W}$ is compact. Therefore $\phi(S \cap \overline{W})$ is compact as well, and closed in particular. This implies that $E \subseteq \phi(S \cap \overline{W})$ (and, in fact, $E = \phi(S \cap \overline{W})$). If p is a point in $E \backslash \phi(S \cap W)$, then we can write p as $\phi(x)$ for some $x \in S \cap (\overline{W} \backslash W)$. Because $x \notin W$, we have that $\phi(x) = x$, and so $p = x$. However, we also have that $x \in S$, and hence $x \in \mathcal{S}_{j,d}$, since $S \subseteq A \cap \mathcal{S}_{j,d}$. (That is, we are assuming from Proposition 11.13 that S is a quasiminimizer restricted to $A \cap \mathcal{S}_{j,d}$, so that S must be contained in $A \cap \mathcal{S}_{j,d}$, as in Definition 11.2.) This shows that p must lie in $\mathcal{S}_{j,d}$. In other words, we have shown that every point p in $E \backslash \phi(S \cap W)$ must lie in $\mathcal{S}_{j,d}$, which is exactly what we wanted, i.e., $E \backslash \big(\phi(S \cap W) \cup \mathcal{S}_{j,d}\big) = \emptyset$. This proves that $\phi(S_0) = E \backslash \mathcal{S}_{j,d}$, as claimed above.

Because $\phi(S_0) = E \backslash \mathcal{S}_{j,d}$, we can apply (11.19) (and the fact that $\widetilde{\phi} = f \circ \phi$, by construction) to obtain that
$$(11.22) \qquad H^d\big(\widetilde{\phi}(S_0)\big) = H^d\big(f(\phi(S_0))\big) \leq C H^d\big(\phi(S_0)\big).$$
On the other hand, $S \cap \widetilde{W} \subseteq S_0 \cup [(S \cap \widetilde{W}) \backslash S_0]$, and therefore
$$H^d\big(\widetilde{\phi}(S \cap \widetilde{W})\big) \leq H^d\big(\widetilde{\phi}(S_0)\big) + H^d\big(\widetilde{\phi}((S \cap \widetilde{W}) \backslash S_0)\big)$$
$$\leq C H^d\big(\phi(S_0)\big) + H^d\big(\widetilde{\phi}((S \cap \widetilde{W}) \backslash S_0)\big),$$
by (11.22).

To control $H^d\big(\widetilde{\phi}((S \cap \widetilde{W}) \backslash S_0)\big)$, let us check that
$$(S \cap \widetilde{W}) \backslash S_0 \subseteq (S \cap W) \backslash S_0.$$
Suppose to the contrary that there is a point x which lies in $(S \cap \widetilde{W}) \backslash S_0$ but *not* in W. Thus $\phi(x) = x$, and so $\phi(x)$ lies in $A \cap \mathcal{S}_{j,d}$, since $S \subseteq A \cap \mathcal{S}_{j,d}$ (as noted previously). From (11.16) we have that
$$\widetilde{\phi}(x) = f\big(\phi(x)\big) = f(x) = x$$
in this situation, so that x does not lie in $\widetilde{W}$, in contradiction to our original hypothesis. Thus the inclusion above is true, and
$$\widetilde{\phi}\big((S \cap \widetilde{W}) \backslash S_0\big) \subseteq \widetilde{\phi}\big((S \cap W) \backslash S_0\big)$$
holds as well.

However, the definition of S_0 implies that
$$\phi\big((S \cap W) \backslash S_0\big) \subseteq \mathcal{S}_{j,d},$$

and (11.16) ensures that f is equal to the identity mapping on $\mathcal{S}_{j,d}$. This proves that
$$\widetilde{\phi}((S \cap W) \backslash S_0) = \phi((S \cap W) \backslash S_0).$$
Therefore,
$$\widetilde{\phi}((S \cap \widetilde{W}) \backslash S_0) \subseteq \phi((S \cap W) \backslash S_0),$$
and so
$$H^d(\widetilde{\phi}((S \cap \widetilde{W}) \backslash S_0)) \leq H^d(\phi((S \cap W) \backslash S_0)).$$
Putting this into our earlier estimate for $H^d(\widetilde{\phi}(S \cap \widetilde{W}))$ we get that
$$\begin{aligned}(11.23) \qquad H^d(\widetilde{\phi}(S \cap \widetilde{W})) &\leq CH^d(\phi(S_0)) + H^d(\widetilde{\phi}((S \cap \widetilde{W}) \backslash S_0)) \\ &\leq CH^d(\phi(S_0)) + H^d(\phi((S \cap W) \backslash S_0)) \\ &\leq CH^d(\phi(S \cap W)).\end{aligned}$$
Let us combine (11.21) and (11.23), to obtain that
$$H^d(S \cap \widetilde{W}) \leq CkH^d(\phi(S \cap W)).$$
This is not quite what we want, because of the $\widetilde{W}$ on the left side, instead of W.

If $x \in (S \cap W) \backslash \widetilde{W}$, then $\phi(x) \neq x$ but $\widetilde{\phi}(x) = x$. This implies that $f(\phi(x)) \neq \phi(x)$ (since $\widetilde{\phi} = f \circ \phi$), and hence that $\phi(x) \in \mathbf{R}^n \backslash \mathcal{S}_{j,d}$, by (11.16). Thus x actually lies in S_0, and $x = \widetilde{\phi}(x)$ lies in $\widetilde{\phi}(S_0)$. In other words, $(S \cap W) \backslash \widetilde{W}$ is contained in $\widetilde{\phi}(S_0)$, and (11.22) then implies that
$$(11.24) \qquad H^d((S \cap W) \backslash \widetilde{W}) \leq H^d(\widetilde{\phi}(S_0)) \leq CH^d(\phi(S_0)).$$
This permits us to conclude that
$$(11.25) \qquad H^d(S \cap W) \leq 2CkH^d(\phi(S \cap W)),$$
using our earlier bound for $H^d(S \cap \widetilde{W})$ and the fact that $S_0 \subseteq S \cap W$ (by the definition of S_0).

This is exactly what we want, i.e., (11.3) for ϕ with k replaced by a suitable constant $2Ck$. This shows that S is a $(U, 2Ck, \widetilde{\delta})$-quasiminimizer restricted to A, as desired. The proof of Proposition 11.13 is now complete.

REMARK 11.26. In the context of Proposition 11.13, the local structure of quasiminimizers restricted to $A \cap \mathcal{S}_{j,d}$ is always quite simple at the scale of 2^{-j}. Indeed, if R is any d-dimensional cube which is a face a dyadic cube $Q \in \Delta_j$, $Q \subseteq A$, then either $S \cap R = R$ or $H^d(S \cap R) = 0$. Otherwise, one could find a Lipschitz mapping ϕ which retracts $S \cap R$ into ∂R and leaves the rest of $\mathcal{S}_{j,d}$ untouched, which would contradict the restricted quasiminimality of S. When $H^d(S \cap R) = 0$ one can use the same method to "clean up" S inside R in the sense of replacing S with an equivalent set $\widetilde{S}$ for which $\widetilde{S} \cap R \subseteq \partial R$.

Thus, for the purpose of studying quasiminimizers restricted to $A \cap \mathcal{S}_{j,d}$, it is enough to consider sets S which can be realized as a (finite) union of d-dimensional faces of $\mathcal{S}_{j,d}$ together with a set of measure zero contained in $\mathcal{S}_{j,d-1}$. In particular, there are, in effect, only finitely many serious competitors in this case (i.e., only finitely many modulo sets of measure 0). We shall make use of this in Section 12.4.

CHAPTER 12

Applications

In this chapter, we want to explain a way to use the analysis of quasiminimizers to study geometric complexity of general sets. The basic idea is to approximate a given set E by quasiminimizing sets, with uniform bounds on the quasiminimizing constants, and then to use Theorem 2.11 to derive information about the approximations. The existence of the quasiminimizing approximations will be obtained by minimizing suitable functionals, functionals with discontinuous coefficients that one has some freedom in choosing. It is very important that the conclusions of Theorem 2.11 come with quantitative estimates which can survive in the limit of the approximations.

With this approach we are following an argument of Morel and Solimini [22]. Their argument was concerned with the existence, under certain conditions, of an Ahlfors-regular curve Γ which would contain a given 1-dimensional regular set E. They observed that an efficient way to establish the existence of Γ was to look at curves containing E with minimal length. The hypotheses on E involved (seemingly) weaker conditions of good behavior at all scales and locations, and the point was to combine the information at different scales and locations into a single curve Γ. This is often done through slightly complicated constructions of combining various pieces together (see [6, 8], for instance), while the minimization of Morel and Solimini produced a choice of Γ all at once. One still has to verify the good behavior of Γ at arbitrary locations and scales separately, but this is easier to do with a good choice of Γ in hand than to produce Γ directly.

The present applications will be directed towards the general theme that if a set E admits upper bounds for its d-dimensional size and lower bounds for its d-dimensional topology, then it should also enjoy substantial tameness in its geometric complexity. This is the "second principle" discussed in [25]. The use of minimization will have the effect of converting information at fixed scales into good behavior that works at all scales at once, which is then more amenable to established geometric techniques. A similar approach was employed in [9], for sets of codimension 1.

See [17] for a very different (and non-variational) approach to related topics.

Variational methods are quite natural in the context of lower bounds for topology, because the latter can prevent competitors for minimization from degenerating, i.e., from collapsing onto sets of small measure or lower dimension. This will be a basic ingredient here, as in Sections 12.2 and 12.3.

We begin in Section 12.1 with some basic notations, conventions, and so forth, and some technical results about the existence of retractions (needed in order to apply Lemma 11.5, for instance). The actual variational problems (in finite polyhedral contexts) will be discussed in Section 12.4. This leads to a sequence of polyhedral quasiminimizers associated to a given set E, and the limiting behavior of this sequence is treated in Section 12.5. In particular, one obtains conclusions about the existence of subsets of E of definite size and controlled behavior under assumptions of upper bounds for $H^d(E)$ and lower bounds for the d-dimensional topology of E. (See also [25].)

12.1. The initial set-up

Let n be a positive integer. We shall employ the same notation and terminology pertaining to dyadic cubes and dyadic skeleta in $\mathbf{R}^n$ as in Chapter 11, just before Proposition 11.13.

Fix an integer j_0. This will determine a basic scale 2^{-j_0} at which our sets will live. For simplicity one could take $j_0 = 0$, for instance.

Let A be a nonempty subset of $\mathbf{R}^n$ which is a finite union of cubes in Δ_{j_0}. Remember that we always take our cubes to be closed, unless the contrary is explicitly stated, so that A is automatically a compact set. For the present purposes one should think of A as not being convex or contractable or anything like that, but carrying instead some nontrivial topology. For example, A might be a cubical approximation to a closed d-dimensional submanifold, like a d-dimensional sphere.

Let Y denote the union of the cubes in Δ_{j_0+2} which intersect A, and let U denote the interior of Y. Thus U is an open subset of $\mathbf{R}^n$ which contains A, and

(12.1) $$\operatorname{dist}(A, \mathbf{R}^n \backslash U) = 2^{-j_0-2}.$$

Roughly speaking, Y looks like A together with a kind of strip that goes around the boundary of A. By defining Y in terms of cubes in Δ_{j_0+2}, we have been careful to make the width of this strip small enough so that we do not accidentally fill in any holes in A, or have some parts of the strip get too close to other parts. This is reflected in the next lemma, which says that A is a retract of Y.

LEMMA 12.2. *There is a mapping $g : Y \to A$ which is a retract (so that $g(x) = x$ for all $x \in A$) and which is Lipschitz with norm bounded by a constant that depends only on n.*

This lemma is similar in spirit to the constructions in Chapter 3, and also to "regular neighborhood" constructions in topology. We shall use this to show that the hypotheses of Lemma 11.5 are satisfied for A and U as defined above. This will permit us to say that quasiminimizers in U restricted to A (in the sense of Definition 11.2) are ordinary quasiminimizers later on, in the proof of Proposition 12.101 in Section 12.4.

Let us begin with some simple facts about the local behavior of Y and A. Let T be a cube in Δ_{j_0} that intersects A but is not contained in A. Set $F = T \cap A$. This is the same as $\partial T \cap A$, since A is a union of cubes in Δ_{j_0}, and T is not contained in A. More precisely, F is a union of (whole) faces of T, with the dimensions of these faces potentially taking any values between 0 and $n-1$ (depending on the relative positions of T and A). Thus if F contains a point p in a k-dimensional face D of T, and if p does not lie in a $(k-1)$-dimensional face of T, then F must contain all of D.

The definition of Y implies that $Y \cap T$ is the union of the cubes in Δ_{j_0+2} which intersect F. This union does not include any of the elements of Δ_{j_0+2} which contain the center of T, since $F \subseteq \partial T$ (and because we were careful to use $j_0 + 2$ rather than $j_0 + 1$). This is a key point, because it permits us to push $Y \cap T$ continuously to the boundary of T without difficulty.

Let $c(T)$ denote the center of T, and let us agree to do the "pushing" along rays that emanate from $c(T)$. This brings us to a second useful point, which is that when we push elements of $Y \cap T$ along rays emanating from $c(T)$ towards the boundary of T, we do not leave $Y \cap T$.

To make this more precise, let Q be a cube in Δ_{j_0+2} which is contained in $Y \cap T$, and which therefore intersects the boundary of T, as mentioned above. It is not true that the radial projection of Q from $c(T)$ to ∂T necessarily stays in Q itself; when $n = 2$, for instance, Q might not be in one of the four corners of T, but it still has a point which is mapped to a corner of T by the radial projection.

In general, pushing points along the rays emanating from $c(T)$ towards the boundary of T has the effect of moving points in Q closer to the vertices of T, or to other lower-dimensional faces of T. In other words, the coordinates of a point x in T are moved to more extreme values by the radial flow, "extreme" for elements of T. One can see this explicitly as follows.

Let x_i denote the ith coordinate of x, $1 \leq i \leq n$, and let b_i, d_i denote the minimum and maximum values of the ith coordinate of elements of T, respectively. Thus T is really the set of $z \in \mathbf{R}^n$ such that $b_i \leq z_i \leq d_i$ for $i = 1, \ldots, n$, and $c(T)$ is described by $c_i(T) = (b_i + d_i)/2$.

If $x_i = c_i(T)$ for some i, then flowing x towards the boundary of T along the ray emanating from $c(T)$ does not change the ith coordinate of x. If $x_i > c_i(T)$, then the ith coordinate becomes larger under the flow, and if $x_i < c_i(T)$, then the ith coordinate becomes smaller. This is easy to verify.

A point $x \in T$ lies in Y exactly when there is an $a \in A \cap T$ such that

(12.3) $$|x_i - a_i| \leq 2^{-j_0-2}$$

for $1 \leq i \leq n$. This is not hard to check, using the assumption that A is a union of cubes in Δ_{j_0}.

We have already mentioned that $F = \partial T \cap A$ consists of entire subfaces of T, perhaps of various dimensions. We can reformulate this as saying that if a lies in $\partial T \cap A$, if $1 \leq i \leq n$, and if $b_i < a_i < d_i$, then one is free to change a_i within the range $[b_i, d_i]$ and stay in $\partial T \cap A$. In particular, one is free to make the coordinates of a more extreme (relative to T) and stay in $\partial T \cap A$.

Using these observations, one can check that the radial flow from the center of T to the boundary of T maps $Y \cap T$ into itself. That is, if an a_i in (12.3) is already at an extreme value, the radial flow will move x_i closer to it. (The extreme value has to be on the same side of $c_i(T)$ as x_i, as one can verify.) If a_i is not at an extreme value, then one can move it freely anyway, as above, and therefore follow x_i as x moves along the ray.

These radial flows provide the first step in the construction of g. Specifically, one can think of defining a Lipschitz mapping $g_1 : Y \to Y$ by taking $x \in Y \cap T$ and flowing it into ∂T using the radial projection from $c(T)$ as above. Thus $g_1(x) \in Y \cap \partial T$ in this case, by the observations above. When $x \in A$ one simply sets $g_1(x) = x$. These local choices are all consistent with each other – agreeing on the overlaps – because the radial projection from $Y \cap T$ into $Y \cap \partial T$ fixes every element of ∂T. Thus we get a well-defined mapping g_1 on all of Y, and it is easy to see that it is Lipschitz with bounded norm. (This uses the fact that points in $Y \cap T$ never get close to the center of T. To get a bound for the Lipschitz constant that does not depend on A, it is helpful to notice that $g(Q \cap Y) \subseteq Q$ for all $Q \in \Delta_{j_0}$. This ensures that the Lipschitz bound depends only on "local" considerations.)

This first step does not normally give a retraction onto A, but only clears out Y from the interiors of the cubes $T \in \Delta_{j_0}$ which are not contained in A. If x is in Y, then $g_1(x)$ will lie in the union of A with

$$(12.4) \qquad \bigcup \{\partial T \cap Y : T \in \Delta_{j_0}, T \subseteq Y\}.$$

To get rid of these extraneous pieces of $(n-1)$-dimensional cubes, we apply exactly the same kind of procedure inside of them. Given an $(n-1)$-dimensional face Z of a cube $T \in \Delta_{j_0}$, either it is completely contained in A, in which case we leave the elements of Z fixed, or it is not. If not, then the middle portion of Z is not at all in Y, for the same reasons as before, and in particular its center does not lie in Y. We can then use radial flows to push $Z \cap Y$ into the boundary of Z, which is of dimension $n-2$. These radial flows take points in Y to points in Y, for exactly the same reasons as before. One simply has 1 fewer coordinate to worry about. (As above, to test for membership in $Y \cap Z$ or $Y \cap \partial Z$, one only has to take $A \cap Z$ into account, and not the rest of A. One can do this at the level of coordinates, as in (12.3). Also, one again has that $A \cap Z$ consists of whole faces of Z, just as for $A \cap T$.)

We now define a mapping g_2 on the union of A with (12.4) by taking g_2 to be the identity on A and using the radial projection from $Z \cap Y$ into $\partial Z \cap Y$ for each Z as in the preceding paragraph. The radial projections automatically fix elements of ∂Z, so that these local choices agree on the overlaps. The values of g_2 lie in

$$(12.5) \qquad A \cup (\mathcal{S}_{j_0, n-2} \cap Y),$$

by construction, and g_2 is Lipschitz with a good bound just as before.

We repeat this $n-2$ more times, to get Lipschitz mappings $g_3, \ldots, g_n$, with

$$(12.6) \qquad g_k : A \cup (\mathcal{S}_{j_0, n-k+1} \cap Y) \to A \cup (\mathcal{S}_{j_0, n-k} \cap Y).$$

In the end we set

$$(12.7) \qquad g = g_n \circ g_{n-1} \circ \cdots \circ g_2 \circ g_1.$$

By construction, g is Lipschitz, with controlled norm, and $g(x) = x$ whenever $x \in A$, because of the corresponding properties for each g_k. We also have that g maps Y into the union of A with

$$(12.8) \qquad \mathcal{S}_{j_0, 0} \cap Y.$$

In other words, (12.8) is the set of points in Y which are vertices of cubes in Δ_{j_0}. It is not hard to see from the definition of Y that these points must lie in A, so that g actually takes values in A. This uses the fact that A is a union of cubes in Δ_{j_0}.

This completes the proof of Lemma 12.2.

Let us record another property of the mapping $g : Y \to A$ produced above. There is a constant C_0 (depending only on n) so that

$$(12.9) \qquad |g(x) - x| \le C_0 \operatorname{dist}(x, A)$$

for each $x \in Y$. This is a quantitative version of the fact that $g(x) = x$ for all $x \in A$. To prove it, let $x \in Y$ be given, and choose $p \in A$ so that

$$(12.10) \qquad |x - p| = \operatorname{dist}(x, A).$$

Then $g(p) = p$, since $p \in A$, and so

$$
\begin{aligned}
(12.11) \qquad |g(x) - x| &= |(g(x) - g(p)) - (x - p)| \\
&\leq |g(x) - g(p)| + |x - p|.
\end{aligned}
$$

The right-hand side is bounded by a constant multiple of $|x-p|$, since g is Lipschitz, and then (12.9) follows from this and (12.10).

We are going to need to have a deformation between g and the identity mapping with good behavior. Define $G : Y \times [0, \infty) \to \mathbf{R}^n$ by the formula

$$(12.12) \qquad G(x,t) = x + \gamma(x,t)\, (g(x) - x),$$

where

$$(12.13) \qquad \gamma(x,t) = \min\left(1, \frac{t}{\mathrm{dist}(x, A)}\right).$$

When $x \in A$, we interpret $\gamma(x,t)$ as being equal to 1. (This does not really matter, since $g(x) - x$ is then 0 anyway.)

LEMMA 12.14. $G(x,t) = x$ for all $x \in A$ and $t \in [0, \infty)$, and when $x \in Y$ and $t = 0$. Also, $G(x,t) \in A$ when $x \in Y$ and $t \geq \mathrm{dist}(x, A)$.

This follows from the definition of G and the corresponding properties for g.

LEMMA 12.15. *The mapping G defined above is Lipschitz on $Y \times [0, \infty)$, with norm bounded by a constant that depends only on n.*

This is a fairly straightforward calculation, but let us go through the main points. If we fix $x \in Y$ and look only at the dependence on t for the moment, then we clearly have that $\gamma(x,t)$ is Lipschitz in t with constant $\leq \mathrm{dist}(x, A)^{-1}$. This implies that $G(x,t)$ is Lipschitz in t with bounded constant, because of (12.9). (When $x \in A$, $\gamma(x,t)$ and $G(x,t)$ are constant in t, and there is nothing to do.)

Now let us look at the x-dependence of G. Notice first that

$$(12.16) \qquad |G(x,t) - x| \leq |g(x) - x| \leq C_0\, \mathrm{dist}(x, A),$$

by (12.9). Thus there is no problem with the behavior of $G(x,t)$ as one approaches A in Y. To check the Lipschitz bound for $G(x,t)$ in x, it suffices to estimate

$$(12.17) \qquad |G(x,t) - G(y,t)|$$

when $x, y \in Y \setminus A$ satisfy

$$(12.18) \qquad |x - y| \leq \frac{1}{2} \min(\mathrm{dist}(x, A), \mathrm{dist}(y, A)).$$

That is, the other cases can be handled directly from (12.16), and bounding (12.17) crudely by

$$(12.19) \qquad |G(x,t) - x| + |x - y| + |G(y,t) - y|.$$

When (12.18) holds, it is not hard to check that

$$(12.20) \qquad \frac{1}{2} \mathrm{dist}(x, A) \leq \mathrm{dist}(y, A) \leq 2\, \mathrm{dist}(x, A).$$

We also know that $\mathrm{dist}(x, A)$ is Lipschitz with constant 1, by a standard argument (which would work for any nonempty set A). Using these facts and the properties of g already established, one can compute that (12.17) is bounded from above by a constant multiple of $|x - y|$ in a standard way. We omit the details, and Lemma 12.15 follows.

Although the definition of G makes sense for all $x \in Y$, we shall sometimes wish to restrict ourselves to a set

(12.21) $$Y_0 = \{x \in \mathbf{R}^n : \operatorname{dist}(x, A) \leq \alpha \, 2^{-j_0}\},$$

where α is $\leq 1/4$, and small enough so that the following works.

LEMMA 12.22. *If α is small enough (depending only on n), then there is a compact subset K of U (the interior of Y) such that $G(x,t) \in K$ for all $x \in Y_0$ and $t \in [0, \infty)$.*

This is an easy consequence of (12.16). One can also choose α small enough that $\operatorname{dist}(K, \mathbf{R}^n \backslash) \geq \alpha \, 2^{-j_0}$.

From now on we assume that α has been fixed as in Lemma 12.22. We shall use $G(x,t)$ to prove the following.

LEMMA 12.23. *Let A and U be as defined above, and let $\delta > 0$ be given. Then there exist $C_1 \geq 1$ and $\delta_1 > 0$ so that the hypotheses of Lemma 11.5 are satisfied (concerning the existence of a mapping $h = h_L$ associated to a given set L). Here C_1 depends only on n, and δ_1 depends only on δ, n, and j_0.*

The fact that C_1 need not depend on δ will be useful later on.

Fix a choice of δ, and let $\delta_1 > 0$ be small, to be chosen soon. As in Lemma 11.5, let L be a compact subset of U which intersects A and satisfies $\operatorname{diam} L < \delta_1$.

Set $D = \{x \in \mathbf{R}^n : \operatorname{dist}(x, L) \leq \operatorname{diam} L\}$. Thus $\operatorname{dist}(x, A) \leq 2\delta_1$ for all $x \in D$, since L intersects A and has diameter less than δ_1. We shall require that $\delta_1 \leq \alpha \, 2^{-j_0 - 1}$, so that

(12.24) $$D \subseteq Y_0.$$

Define $f : \mathbf{R}^n \to [0, \infty)$ by

(12.25) $$f(x) = \min(\operatorname{dist}(x, \mathbf{R}^n \backslash D), \operatorname{diam} L).$$

Thus f is Lipschitz with constant 1, and $f(x) = \operatorname{diam} L$ when $x \in L$, by the definition of D. Since L intersects A, we have that $\operatorname{dist}(x, A) \leq \operatorname{diam} L$ for all $x \in L$, and hence

(12.26) $$f(x) \geq \operatorname{dist}(x, A) \qquad \text{when } x \in L.$$

We define $h = h_L : \mathbf{R}^n \to \mathbf{R}^n$ as follows. If $x \in Y$, then we put

(12.27) $$h(x) = G(x, f(x)),$$

where $G(x, t)$ is as in (12.12). When $x \in \mathbf{R}^n \backslash Y$, we set $h(x) = x$. Thus in fact we have that $h(x) = x$ whenever $x \in \mathbf{R}^n \backslash D$, since f vanishes on this set and $G(x, 0) = x$ for all $x \in Y$. In particular, there are no discontinuities of h at the boundary of Y.

If $x \in A$, then $h(x) = x$, since $G(x, t) = x$ for all $t \geq 0$. If $x \in L$, then $f(x) \geq \operatorname{dist}(x, A)$, as in (12.26), so that $h(x) = G(x, f(x))$ lies in A, by Lemma 12.14. This shows that (11.6) and (11.7) are satisfied. One can also check that h is Lipschitz, with a constant that depends only on n, because of the corresponding statement for G (Lemma 12.15), and the fact that f is Lipschitz with constant 1.

This choice of h comes with a natural homotopy $H : \mathbf{R}^n \times [0, 1] \to \mathbf{R}^n$, which is defined by

(12.28) $$H(x, s) = G(x, sf(x))$$

when $x \in Y$, $H(x,s) = x$ when $x \in \mathbf{R}^n \backslash Y$. We actually have that $H(x,s) = x$ for all $x \in \mathbf{R}^n \backslash D$ and $s \in [0,1]$, for the same reason as before. Thus no discontinuities are introduced at the boundary of Y, and H is even Lipschitz. We have that

(12.29) $$H(x,0) = G(x,0) = x$$

for all x, and that $H(x,1) = h(x)$ for all x too.

We need to verify (11.9) and its counterpart for $H(x,s)$. It will be useful to make some auxiliary definitions. Given $t \geq 0$, set

(12.30) $$E(t) = \{z \in \mathbf{R}^n : z = G(x,t) \text{ for some } x \in D\},$$

and put

(12.31) $$E = \bigcup_{t \geq 0} E(t).$$

Let p be an element of $A \cap L$. This exists, by our original assumptions on L. We have that $p \in D$, since $D \supseteq L$, and we also have that $p \in E(t)$ for every t. This is because $G(p,t) = p$ for all t, since $p \in A$.

Let us check that

(12.32) $$E \subseteq \overline{B}(p, 2\, C(n) \operatorname{diam} L) \subseteq B(p, 2\, C(n)\, \delta_1),$$

where $C(n)$ is a bound for the Lipschitz constant for H (which depends only on n). The second inclusion is automatic, because $\operatorname{diam} L < \delta_1$ by assumption. Since $p \in L$, we have that

(12.33) $$D \subseteq \overline{B}(p, 2 \operatorname{diam} L),$$

by the definition of D. From here we get that

(12.34) $$E(t) \subseteq \overline{B}(p, 2\, C(n) \operatorname{diam} L),$$

because $E(t)$ is the image of D under $x \mapsto G(x,t)$ (by construction), and because $G(p,t) = p$. This gives (12.32), since E is just the union of the $E(t)$'s.

We are now ready to choose δ_1. Specifically, we require that

(12.35) $$\delta_1 < \min(\alpha\, 2^{-j_0-1}, 8\, C(n)^{-1}\, \delta),$$

and we do not need to impose any other conditions on δ_1. For the sake of definiteness one can take δ_1 to be one-half the right side of (12.35), for instance.

Under these conditions we have that

(12.36) $$\operatorname{diam} E < \frac{\delta}{2}.$$

That is, the diameter of E is $\leq 4C(n)\delta_1$, by (12.32), and then this is less than $\delta/2$ because of (12.35).

Let W_h denote the set of points $x \in \mathbf{R}^n$ such that $h(x) \neq x$, and let $\widehat{W}(h)$ be the set of points of the form x or $H(x,s)$, where $x \in \mathbf{R}^n$ and $s \in [0,1]$ satisfy $H(x,s) \neq x$. These definitions are the same as in Lemma 11.5. Let us verify that

(12.37) $$W_h \cup h(W_h), \widehat{W}(h) \subseteq E.$$

Notice first that

(12.38) $$h(x) = H(x,s) = x \qquad \text{when } x \notin D,$$

by construction (since $f(x) = 0$ when $x \notin D$). This implies that $W_h \subseteq D$, and hence that $W_h \subseteq E$, since $E(0) = D$ automatically. We also obtain that $h(W_h) \subseteq E$, because of the inclusion $W_h \subseteq D$, (12.30), (12.31), and the definition (12.27) of

h. Similarly, $\widehat{W}(h) \subseteq E$, since (12.38) ensures that the relevant x's lie in D, while the relevant $H(x,s)$'s lie in E because of this, (12.30), (12.31), and the definition (12.28) of H. This shows that (12.37) holds, from which we obtain

$$(12.39) \qquad \operatorname{diam}(W_h \cup h(W_h)), \ \operatorname{diam} \widehat{W}(h) \le \operatorname{diam} E < \frac{\delta}{2},$$

by (12.36).

It remains to check that

$$(12.40) \qquad \operatorname{dist}(W_h \cup h(W_h), \mathbf{R}^n \backslash U) > 0 \quad \text{and} \quad \operatorname{dist}(\widehat{W}(h), \mathbf{R}^n \backslash U) > 0.$$

It suffices to show

$$(12.41) \qquad \operatorname{dist}(E, \mathbf{R}^n \backslash U) > 0,$$

since $W_h \cup h(W_h)$ and $\widehat{W}(h)$ are contained in E. Remember that δ_1 was chosen so that D is contained in Y_0, as in (12.24). (This was also included in (12.35).) Because of this, (12.41) follows from Lemma 12.22 and the definition of E. (I.e., $E \subseteq K$, where K is as in Lemma 12.22.)

To summarize, we have shown that the mapping h defined above enjoys all the properties required in Lemma 11.5, and the proof of Lemma 12.23 is now complete.

Let us mention one other application of Lemma 12.2.

LEMMA 12.42. *Let E be a subset of A, and let $\phi : E \to A$ be a continuous mapping which satisfies*

$$(12.43) \qquad |\phi(x) - x| \le 2^{-j_0 - 2}$$

for all $x \in E$. Then ϕ is homotopic to the identity through mappings of E into A.

To prove this, we start with the homotopy between ϕ and the identity given by

$$(12.44) \qquad \Phi(x,s) = s\,\phi(x) + (1-s)x.$$

This may not take values in A, but it does take values in Y, because of (12.43) and the definition of Y. More precisely,

$$(12.45) \qquad \operatorname{dist}(\Phi(x,s), A) \le |\Phi(x,s) - x| = s|\phi(x) - x| \le 2^{-j_0 - 2}$$

for all $x \in E$ and $s \in [0,1]$, by (12.43), and this implies that $\Phi(x,s) \in Y$. This permits us to compose Φ with the retraction from Y onto A given by Lemma 12.2 to get a homotopy between ϕ and the identity through mappings with values in A, as desired. This proves Lemma 12.42.

12.2. Stability of sets

Fix an integer d, $0 < d < n$, and let A, U, j_0, etc., be as before. We shall be interested in knowing when a subset of A can be deformed into a set which is small in terms of d-dimensional measure, and when it cannot.

PROPOSITION 12.46. *There is a constant $\theta > 0$ (which depends on n and j_0) with the following property. Let E be any compact subset of A. Then either*

$$(12.47) \qquad H^d(f(E)) \ge \theta \quad \textit{for every continuous mapping } f : E \to A$$
$$\textit{which is homotopic to the identity}$$
$$\textit{through mappings from } E \textit{ into } A,$$

or

(12.48)   there is a continuous mapping $\phi : E \to A$ which
maps E into $\mathcal{S}_{j_0,d-1}$, and which is homotopic
to the identity through mappings from E into A.

In other words, either E is "stable", in the sense that one cannot deform it into a set of small measure, or it is not, in which case it can be deformed into a polyhedron of dimension $< d$. Roughly speaking, the stability property is a kind of lower bound for the topology of E. We shall say more about this in Section 12.3.

The constant θ can be taken so that it is a constant that depends only on n times $2^{-j_0 d}$. This is clear from considerations of scaling, and it will be apparent in the proof as well.

One could strengthen (12.47) a bit by saying that for any f as in (12.47) there is a cube Q in Δ_{j_0} such that $H^d(f(E) \cap Q) \geq \theta$. This is a bit better in some situations, for avoiding unnecessary dependence of constants on the diameter or volume of A, for instance. For some purposes it is also better to work with mappings from E to A which can be extended to maps from A to A, with the extension being homotopic to the identity as well. This comes up in Remark 12.171 in Section 12.5, and it can be accommodated through the same argument, but with a few extra steps.

Proposition 12.46 is another version of the usual Federer-Fleming arguments, as in Proposition 3.1 and Lemma 11.14. We shall derive it from the next lemma.

LEMMA 12.49. *There is a $\theta > 0$ (depending on n and j_0) so that if F is a compact subset of A which satisfies*

(12.50) $$H^d(F) < \theta,$$

then there is a Lipschitz mapping $\psi : F \to \mathbf{R}^n$ such that $\psi(F) \subseteq \mathcal{S}_{j_0,d-1}$ and $\psi(F \cap Q) \subseteq Q$ for all $Q \in \Delta_{j_0}$.

As indicated above, one could replace (12.50) with the requirement that

$$H^d(F \cap Q) < \theta \qquad \text{for all } Q \in \Delta_{j_0}.$$

To prove Lemma 12.49, let F be a compact subset of A with $H^d(F) < \infty$, and let $g : \mathbf{R}^n \to \mathbf{R}^n$ be the Lipschitz mapping provided by Lemma 11.14 (with $E = F$ and $j = j_0$). From (11.18) we have that

(12.51) $$g(Q) \subseteq Q \qquad \text{for all } Q \in \Delta_{j_0}.$$

In particular, g maps A into A, since A is a union of cubes in Δ_{j_0}. We also have that $g(F) \subseteq \mathcal{S}_{j_0,d}$, by (11.17), and that

(12.52) $$H^d(g(F)) \leq C_0 H^d(F),$$

from (11.19) and (11.16). This constant C_0 depends only on n, and not on F in particular.

We now choose θ so that

(12.53) $$C_0 \theta = 2^{-j_0 d}.$$

Assuming that $H^d(F) < \theta$, as in the lemma, we get that

(12.54) $$H^d(g(F)) < 2^{-j_0 d}.$$

The right-hand side is the same as the measure of a d-dimensional cube of sidelength 2^{-j_0}. (Actually, with our normalization for Hausdorff measure, the latter statement may be off by a constant factor (that depends only on d), but this can easily be corrected by an adjustment to (12.53).)

Let T be any d-dimensional cube which arises in $\mathcal{S}_{j_0,d}$, i.e., a d-dimensional face of a cube Q in Δ_{j_0}. Then T has sidelength 2^{-j_0}, and

$$(12.55) \qquad H^d(g(F) \cap T) \leq H^d(g(F)) < 2^{-jd} = H^d(T).$$

This implies that there is a point $p(T)$ in the interior of T that does not lie in $g(F)$.

Define $h : g(F) \to \mathcal{S}_{j_0,d-1}$ as follows. Given a d-dimensional cube T in $\mathcal{S}_{j_0,d}$ as above, choose h on $g(F) \cap T$ so that it agrees with the radial projection of $T \setminus \{p(T)\}$ into ∂T centered at $p(T)$. We do this for every such T. This defines h on all of $g(F)$, since $g(F) \subseteq \mathcal{S}_{j_0,d}$. These local definitions are consistent with each other, because they all agree with the identity on the boundaries of the T's. Thus h is continuous, and even Lipschitz (since $g(F) \cap T$ is compact and does not contain $p(T)$).

Now set $\psi = h \circ g$. This is Lipschitz, because g and h are, and it maps F into $\mathcal{S}_{j_0,d-1}$. It also satisfies $\psi(F \cap Q) \subseteq Q$ for all $Q \in \Delta_{j_0}$, because of the corresponding properties for g and h. This completes the proof of Lemma 12.49.

Notice that $\psi(F)$ is automatically contained in A, since $\psi(F \cap Q) \subseteq Q$ for all $Q \in \Delta_{j_0}$, and A is a union of cubes in Δ_{j_0}. We also have that $\psi : F \to A$ is homotopic to the identity through mappings of F into A. For this we can even use the "trivial" homotopy

$$(12.56) \qquad \Psi(x,t) = t\,\psi(x) + (1-t)\,x, \qquad 0 \leq t \leq 1.$$

Indeed, let $x \in F$ and $t \in [0,1]$ be given, and choose $Q \in \Delta_{j_0}$ so that $x \in Q$ and $Q \subseteq A$. Then $\psi(x)$ lies in Q too, and this ensures that $\Psi(x,t) \in Q$ for each t, because Q is convex. In particular, $\Psi(x,t)$ lies in A for all t.

It remains to prove Proposition 12.46. Let $\theta > 0$ be as in Lemma 12.49, and let E be an arbitrary compact subset of A. Assume that (12.47) is not true, so that there is a continuous mapping $f : E \to A$ which satisfies $H^d(f(E)) < \theta$ and is homotopic to the identity through mappings from E into A. Apply Lemma 12.49 to $F = f(E)$ to get a mapping $\psi : F \to A$ as above, and set $\phi = \psi \circ f$. Thus ϕ is a continuous mapping that maps E into $\mathcal{S}_{j_0,d-1}$. It is not hard to check that ϕ is homotopic to the identity through mappings from E into A, using the corresponding homotopies for f and ψ. (Specifically, $\Psi(f(x),t)$ gives a homotopy from f to ϕ which can be combined with the one between f and the identity that exists by assumption.) This completes the proof of Proposition 12.46.

Let us mention a simple fact, which we shall use later on.

LEMMA 12.57. *Suppose that E is a compact subset of A which satisfies the stability condition (12.47). If $g : E \to A$ is a continuous mapping which is homotopic to the identity through mappings of E into A, then $\widetilde{E} = g(E)$ also satisfies (12.47).*

This is an easy consequence of the definitions. The main point is that if $\beta : g(E) \to A$ is a continuous mapping which is homotopic to the identity through mappings from $g(E)$ into A, then $\beta \circ g$ is homotopic to the identity through mappings from E to A, because of the corresponding assumption on g.

For the rest of this section, we shall discuss a strengthening of the stability condition (12.47) that we shall need, and which is quite natural in its own right. The main point is that instead of having lower bounds for size in terms of Hausdorff *measure*, one can use Hausdorff *content*. We begin by recalling the definition of the latter.

Let E be a subset of $\mathbf{R}^n$. The *d-dimensional Hausdorff content* of E, denoted $H^d_{con}(E)$, is defined by

$$(12.58) \qquad H^d_{con}(E) = \inf\left\{\sum_\ell (\operatorname{diam} Y_\ell)^d : \{Y_\ell\} \text{ is a sequence of sets in } \mathbf{R}^n \text{ which covers } E\right\}.$$

This is the same as $H^d_\delta(E)$, as defined in (0.1), with $\delta = \infty$.

From the definitions (12.58), (0.2) of Hausdorff content and Hausdorff measure we have that

$$(12.59) \qquad H^d_{con}(E) \leq H^d(E)$$

for any set E. More precisely, $H^d_{con}(E)$ and $H^d(E)$ are both defined through the same kind of sum associated to coverings of E, but Hausdorff measure $H^d(E)$ uses more restrictive coverings. Hausdorff measure uses coverings $\{Y_\ell\}$ in which the diameters of the Y_ℓ's are less than any positive number δ that is given in advance, while Hausdorff content allows arbitrary coverings, without conditions on the diameters of the sets. For instance, the Hausdorff content of a bounded set E is always finite, and less than or equal to $(\operatorname{diam} E)^d$, while the Hausdorff measure can easily be infinite (as in the case of sets of larger dimension).

Although (12.59) cannot in general be reversed, it is true that

$$(12.60) \qquad H^d_{con}(E) = 0 \quad \text{if and only if} \quad H^d(E) = 0.$$

This is a standard observation that is easy to check. The point is that if $H^d_{con}(E)$ is equal to 0, then the coverings $\{Y_\ell\}$ which give arbitrarily small values to the sums $\sum_\ell (\operatorname{diam} Y_\ell)^d$ also have arbitrarily small values for $\sup_\ell \operatorname{diam} Y_\ell$.

Note that $H^d(E) \leq C H^d_{con}(E)$ is true when E is contained in a set which is Ahlfors-regular of dimension d, and when E is a subset of a d-plane in particular. This is not hard to verify.

It is easy to see that the Hausdorff content enjoys the usual countable subadditivity properties, i.e., the content of a countable union is less than or equal to the sum of the contents of the individual sets. However, *additivity* for disjoint sets does not work in general. This is one of the main reasons that one normally works with Hausdorff measure instead of Hausdorff content. On the other hand, Hausdorff content has the nice feature of being much more stable with respect to limits in the Hausdorff metric than Hausdorff measure is. We shall return to this in Section 12.5 (especially Proposition 12.116).

For now we want to look at the way that the stability of a set implies a lower bound on Hausdorff content, and not just Hausdorff measure, as in (12.47). This is very natural, because the condition (12.48) which is the opposite of (12.47) already cooperates with approximations in the Hausdorff metric, in a way which is not properly reflected in (12.47) (and fits better with Hausdorff content).

PROPOSITION 12.61. *Let A, U, j_0, etc., be as before. There is a constant $\theta' > 0$, which depends only on n and j_0, so that the following is true. Let F be a compact subset of A such that*

$$(12.62) \qquad H^d_{con}(F \cap Q) < \theta' \qquad \text{for all } Q \in \Delta_{j_0}.$$

Then there is a Lipschitz mapping $\psi' : F \to A$ such that $\psi'(F) \subseteq \mathcal{S}_{j_0,d-1}$ and $\psi'(F \cap Q) \subseteq Q$ for all $Q \in \Delta_{j_0}$. Also, ψ' is homotopic to the identity through mappings from F into A.

In particular, the stability condition (12.47) implies a lower bound for Hausdorff content in terms of θ' (since the conclusions of Proposition 12.61 imply that (12.48) holds). One can also formulate this in terms of a dichotomy, in the same manner as in Proposition 12.46.

The constant θ' in Proposition 12.61 can be taken to be a constant that depends only on n times $2^{-j_0 d}$, just as before. This will be clear from the proof too.

The proof of Proposition 12.61 is very similar to that of Lemma 12.49 and Proposition 12.46, and it relies on constructions like the ones in Lemma 11.14 and Proposition 3.1. The non-additivity of Hausdorff content does make for a bit of trouble, however. This comes up in the proof of Lemma 3.22, the inequality (3.24) especially, and this is the first matter that we want to address.

We begin by setting some notation. Let T be an m-dimensional cube in $\mathbf{R}^m$ with sidelength 2^{-j_0}. Let $\frac{1}{2}T$ denote the cube which has the same center as T and half the side length of T. (Of course the sides of $\frac{1}{2}T$ should be parallel to the sides of T as well.) Given a point $\xi \in \frac{1}{2}T$, let

(12.63) $$\sigma_{\xi,T} : T\backslash\{\xi\} \to \partial T$$

denote the usual "radial projection" centered at ξ. This is the same as the mapping $\theta_{\xi,T}$ in (3.18), but we want to use a different name for it here.

The following provides a substitute for Lemma 3.22 and the inequality (3.24), with Hausdorff content in place of Hausdorff measure.

LEMMA 12.64. *There exist positive constants C, C', and η, depending only on n, so that the following is true. Let T and $\sigma_{\xi,T} : T\backslash\{\xi\} \to \partial T$ be as above, and assume that $d < m \leq n$ (where d, m, and n are all positive integers). Let W be a compact subset of T, and assume also that*

(12.65) $$H^d_{con}(W) < \eta\, 2^{-j_0 d}.$$

As in the definition (12.58) of Hausdorff content, this means that there is a sequence of sets $\{Y_\ell\}_\ell$ such that

(12.66) $$W \subseteq \bigcup_\ell Y_\ell \quad \text{and} \quad \sum_\ell (\operatorname{diam} Y_\ell)^d < \eta\, 2^{-j_0 d}.$$

(One can also require that $\sum_\ell (\operatorname{diam} Y_\ell)^d$ be as close to $H^d_{con}(W)$ as one wants.) Under these conditions, there is a measurable set $X \subseteq \frac{1}{2}T$ such that

(12.67) $$W \cap \tfrac{1}{2}T \subseteq X,$$

the m-dimensional Lebesgue measure of X is at most one-half the measure of $\frac{1}{2}T$, and

(12.68) $$H^d_{con}(\sigma_{\xi,T}(W)) \leq C \sum_\ell \frac{(\operatorname{diam} T)^d}{(\operatorname{dist}(\xi, Y_\ell) + \operatorname{diam} Y_\ell)^d} (\operatorname{diam} Y_\ell)^d$$

for all $\xi \in \frac{1}{2}T \backslash X$.

Furthermore,

(12.69) $$\frac{1}{|\frac{1}{2}T\backslash X|} \int_{\xi \in \frac{1}{2}T\backslash X} H^d_{con}(\sigma_{\xi,T}(W))\, d\xi \leq C' \sum_\ell (\operatorname{diam} Y_\ell)^d < C'\eta\, 2^{-j_0 d}$$

(where $|\frac{1}{2}T\backslash X|$ denotes the Lebesgue measure of $\frac{1}{2}T\backslash X$, and the integral on the left side is taken with respect to Lebesgue measure too), and there exists $\xi \in \frac{1}{2}T\backslash X$ so that

(12.70) $$H_{con}^d(\sigma_{\xi,T}(W)) \leq C' \sum_\ell (\operatorname{diam} Y_\ell)^d < C'\eta\, 2^{-j_0 d}.$$

As in Lemma 3.22, one does not really need to even think about measurability issues for the integral in (12.69), because the proof will show that the inequality holds with the integral interpreted as an outer integral, which is fine for our purposes. (This is not to say that measurability would be too complicated anyway.)

The proof of the lemma is pretty straightforward, once one has everything set up properly, as above. Let W be given, with the condition (12.65) on its Hausdorff content, and let $\{Y_\ell\}_\ell$ be as in (12.66). We may as well assume that each Y_ℓ is contained in T, since $W \subseteq T$. (Otherwise one can simply replace Y_ℓ with $Y_\ell \cap T$.) For each ℓ, define a relatively open set $\widehat{Y}_\ell$ in $\frac{1}{2}T$ by

(12.71) $$\widehat{Y}_\ell = \{x \in \tfrac{1}{2}T : \operatorname{dist}(x, Y_\ell) < \operatorname{diam} Y_\ell\}.$$

With this choice we have that $Y_\ell \cap \frac{1}{2}T \subseteq \widehat{Y}_\ell$, and

(12.72) $$\operatorname{diam} \widehat{Y}_\ell \leq \min(3 \operatorname{diam} Y_\ell, \operatorname{diam} \tfrac{1}{2}T)$$

for all ℓ.

Define X by

(12.73) $$X = \bigcup_\ell \widehat{Y}_\ell.$$

Thus $W \cap \frac{1}{2}T \subseteq X$, as required in (12.67), because of (12.66) and the fact that $\widehat{Y}_\ell \supseteq Y_\ell \cap \frac{1}{2}T$ for all ℓ. Writing $|X|$ for the m-dimensional Lebesgue measure of X, we have that

(12.74) $$|X| \leq \sum_\ell |\widehat{Y}_\ell| \leq v_m \sum_\ell (\operatorname{diam} \widehat{Y}_\ell)^m,$$

where v_m is a constant that depends only on m. This simplifies to

(12.75) $$\begin{aligned}|X| &\leq v_m\, 3^d\, (\operatorname{diam} \tfrac{1}{2}T)^{m-d} \sum_\ell (\operatorname{diam} Y_\ell)^d \\ &< v_m\, 3^d\, (\operatorname{diam} \tfrac{1}{2}T)^{m-d}\, \eta\, 2^{-j_0 d},\end{aligned}$$

by (12.72) and (12.66). If η is small enough, in a way that depends only on n, then (12.75) implies that

(12.76) $$|X| \leq \frac{1}{2}|\tfrac{1}{2}T|.$$

This is the only condition that we need to impose on η.

Now let us check (12.68). Thus we want to have a bound for $H_{con}^d(\sigma_{\xi,T}(W))$ when $\xi \in \frac{1}{2}T\backslash X$. Note that $W \subseteq T\backslash\{\xi\}$ when $\xi \in \frac{1}{2}T\backslash X$, because of (12.67). For that matter, we also have that $Y_\ell \subseteq T\backslash\{\xi\}$ when $\xi \in \frac{1}{2}T\backslash X$, by the definitions (12.71) and (12.73) of $\widehat{Y}_\ell$ and X.

Consider the sequence of sets $\{\sigma_{\xi,T}(Y_\ell)\}_\ell$. These sets cover $\sigma_{\xi,T}(W)$, since the Y_ℓ's cover W. Therefore

(12.77) $$H_{con}^d(\sigma_{\xi,T}(W)) \leq \sum_\ell (\operatorname{diam} \sigma_{\xi,T}(Y_\ell))^d.$$

On the other hand,

(12.78) $$\operatorname{dist}(\xi, Y_\ell) \geq \operatorname{diam} Y_\ell \qquad \text{for all } \ell$$

when $\xi \in \frac{1}{2}T \backslash X$, because of the definitions (12.73) and (12.71) of X and $\widehat{Y}_\ell$. This implies that

(12.79) $$\operatorname{diam} \sigma_{\xi, T}(Y_\ell) \leq C \frac{\operatorname{diam} T}{\operatorname{dist}(\xi, Y_\ell) + \operatorname{diam} Y_\ell} \operatorname{diam} Y_\ell \qquad \text{for all } \ell$$

when $\xi \in \frac{1}{2}T \backslash X$, where C is a constant that depends only on m. This is not hard to check, using the definition of $\sigma_{\xi, T}$ as the radial projection from $T \backslash \{\xi\}$ to ∂T.

Combining (12.77) and (12.79), we get (12.68). Let us use this to obtain the integral estimate (12.69). The first main point is that

(12.80) $$\int_{\frac{1}{2}T} \frac{(\operatorname{diam} T)^d}{(\operatorname{dist}(\xi, Y_\ell) + \operatorname{diam} Y_\ell)^d} \, d\xi \leq C_1 (\operatorname{diam} T)^m \qquad \text{for all } \ell,$$

where C_1 is a constant that depends only on m and d. This is not hard to check, using the fact that $d < m$. One should not worry too much about the factors of $\operatorname{diam} T$, as they are determined by considerations of scaling. In other words, one may as well check (12.80) when T is the unit cube. Then (12.80) is a simple integral estimate, which in fact can be derived from the special case where Y_ℓ consists of a single point.

Once one has (12.80), one can throw in the factor of $1/|\frac{1}{2}T \backslash X|$ to get

(12.81) $$\frac{1}{|\frac{1}{2}T \backslash X|} \int_{\frac{1}{2}T} \frac{(\operatorname{diam} T)^d}{(\operatorname{dist}(\xi, Y_\ell) + \operatorname{diam} Y_\ell)^d} \, d\xi \leq C' \qquad \text{for all } \ell,$$

where C' is a constant that depends only on m. This uses (12.76), which ensures that $|\frac{1}{2}T \backslash X| \geq \frac{1}{2}|\frac{1}{2}T|$. Now we can sum in ℓ and apply Fubini's theorem to obtain

(12.82) $$\frac{1}{|\frac{1}{2}T \backslash X|} \int_{\frac{1}{2}T} \sum_\ell \frac{(\operatorname{diam} T)^d}{(\operatorname{dist}(\xi, Y_\ell) + \operatorname{diam} Y_\ell)^d} (\operatorname{diam} Y_\ell)^d \, d\xi \leq C' \sum_\ell (\operatorname{diam} Y_\ell)^d$$

from (12.81). This and (12.68) imply (12.69). (Note that the last inequality in (12.69) follows from (12.66).)

The last part of Lemma 12.64, concerning the existence of $\xi \in \frac{1}{2}T \backslash X$ so that (12.70) holds, is a standard consequence of (12.69). This completes the proof of Lemma 12.64.

Lemma 12.64 is very similar to Lemma 3.22, in the kind of averaging argument which is used. The main difference is that Hausdorff content is not as local as Hausdorff measure, which meant that we could not write down an estimate for the Hausdorff content $H^d_{con}(\sigma_{\xi, T}(W))$ in terms of a sum or integral as directly as before, in (3.24). (Compare also with (3.20).) Instead we had (12.68), which did not apply to all points $\xi \in \frac{1}{2}T$, but which worked just as well by the end.

Note that if the Hausdorff content of W were not small compared to $2^{-j_0 d}$, as in the assumption (12.65), then one would not want to say that the Hausdorff content of the radial projection is small anyway. In that event one could use the trivial estimate, that the d-dimensional Hausdorff content of the radial projection of W is less than or equal to the Hausdorff content of ∂T, which is $O(2^{-j_0 d})$ (even if ∂T has dimension larger than d). This estimate would be about as much as one could expect. In other words, the hypothesis (12.65) on the Hausdorff content of

W is not really a serious restriction for the final conclusions about the Hausdorff content of the radial projections. (For Hausdorff *measure* the situation is somewhat different, because bounds for d-dimensional Hausdorff measure contain nontrivial information for subsets of cubes of larger dimension, even when the measure is large. Compare with Lemma 11.19. By contrast, the Hausdorff content of any bounded set is finite, no matter the dimension, and controlled in terms of the diameter.)

Once one has Lemma 12.64, Proposition 12.61 can be established in nearly the same manner as Lemma 12.49 and Proposition 12.46 were, using Federer-Fleming arguments, as in Proposition 3.1 and Lemma 11.14. More precisely, one starts by making radial projections of the given set F inside of each cube Q in Δ_{j_0} which is also contained in A. One chooses these radial projections using Lemma 12.64, and in this way one can push F into the boundaries of these cubes. Assuming that $d < n - 1$, one repeats the process in the $(n-1)$-dimensional cubes which make up the faces of the cubes Q before, and then in cubes of lower dimension.

At each step one would like to make sure that there is not too much d-dimensional Hausdorff content of F and its successive images inside of any given m-dimensional cube, as in the hypothesis (12.65) in Lemma 12.64. This comes from the hypothesis (12.62) in Proposition 12.61, at least if θ' is small enough. One should be a bit careful about this, in that this is supposed to work not only for the first layer of the construction, working inside of the cubes of the top dimension n (and in parallel), but also in succeeding steps when one makes projections into lower dimensional skeleta. However, it is clear from Lemma 12.64 and its proof that we get to choose the radial projections at each step so that they only increase the Hausdorff content by a bounded factor. This corresponds to the fact that one is free to choose the covering $\{Y_\ell\}_\ell$ in Lemma 12.64 so that $\sum_\ell (\operatorname{diam} Y_\ell)^d$ is as close to the Hausdorff content of W as one wants. If the Hausdorff content ever becomes 0, then the Hausdorff measure vanishes too, as in (12.60), and one is back to the earlier situation anyway.

One should be a bit careful about another point as well, which is that in later generations of the construction (i.e., after the first one), the sets to which Lemma 12.64 and the projections are applied do not (in general) come from $F \cap Q$ for a single $Q \in \Delta_{j_0}$. At the first stage one simply projects $F \cap Q$ to the boundary of Q, but in the next stage one makes projections in each of the individual faces of dimension $n-1$, and these faces can receive contributions from more than one cube Q. However, it is only the cubes Q which contain the given face that contribute, and the number of these is bounded. This also works in later stages, with faces of lower dimension.

Thus one starts with the hypothesis (12.62) in Proposition 12.61, with the Hausdorff content of $F \cap Q$ being bounded by θ' for each $Q \in \Delta_{j_0}$. With each step in the construction, the "local" Hausdorff contents of the sets in question (contained in faces of the cubes Q of some dimension) can increase by a bounded factor. The bounds for these factors are known in advance, from Lemma 12.64, and depend only on the dimension. This permits us to be sure that we can choose θ' small enough at the beginning, so that at all successive stages of the construction the local Hausdorff contents are small enough so as to be compatible with the hypothesis (12.65) in Lemma 12.64. Keep in mind that the number of "stages" in the construction is bounded here, by the dimension n.

Once one has this, the rest of the argument is nearly the same as before, in the proof of Lemma 12.49. One first gets a mapping $g : F \to A$ which is Lipschitz and pushes F into the d-skeleton of A. One does this in such a way that $g(Q) \subseteq Q$ for all $Q \in \Delta_{j_0}$ which are contained in A, as in (12.51), and so that the Hausdorff content of $g(F) \cap Q$ is bounded by a constant multiple of the Hausdorff content of $F \cap Q$ for all $Q \in \Delta_{j_0}$, with a constant that depends only on n. The latter reflects a culmination of the bounds discussed in the preceding paragraphs. If the initial choice of θ' is small enough, then this ensures that $g(F)$ does not completely cover any of the d-dimensional faces of any of the cubes $Q \in \Delta_{j_0}$. This permits one to make radial projections again inside the d-dimensional faces, as with the mapping h near the end of the proof of Lemma 12.49, to push $g(F)$ into the $(d-1)$-dimensional skeleton by a Lipschitz mapping in the same manner as before. (For this one does not need to be careful about the way that the individual radial projections are chosen, just that they are centered at points which avoid $g(F)$. There is no longer a question of bounds on measure or Hausdorff content – since the images are now going into cubes of dimension $d-1$ anyway – and no longer any need for choosing ξ's through averaging arguments, as before.)

From here one can get a mapping $\psi' : F \to A$ with the properties required in Proposition 12.61, for the same reasons as in the earlier situation.

In short, while the "local" choices of radial projections have to be carried out a bit differently (at least for the steps down to dimension d), as in Lemma 12.64, the way that these local choices are used in the construction is practically the same as before. One has to be slightly careful, for the way that the conditions under which one can make these choices of radial projections changes (small Hausdorff content instead of finite Hausdorff measure), but this does not cause trouble here, because one can accommodate the extra restrictions by taking θ' small enough in Proposition 12.61. For that matter, there is not too much to say when the Hausdorff content is not reasonably small anyway, since a cube of sidelength 2^{-j_0} has d-dimensional Hausdorff content which is $O(2^{-j_0 d})$, even if the cube itself has dimension larger than n.

This completes our discussion of the proof of Proposition 12.61.

12.3. Topological interpretations

One can think of the stability condition (12.47) as a kind of topological nondegeneracy condition. If one restricts oneself to mappings f which are small perturbations of the identity (for which there are analogous dichotomies as in Proposition 12.46), then this type of stability is close in spirit to the requirement that a set have topological dimension $\geq d$. See [16] concerning topological dimension theory, including the relationship between dimension and measure (Theorem VII 2 on p104 of [16]), Alexandroff's theorem on topological dimension and approximation by finite polyhedra (p72 in [16]), and results on stable values (Section VI.1 of [16]).

We shall not pursue this general theme here, but we would like to illustrate it with some examples of topological conditions which imply (12.47). We begin with *linking* conditions.

Let d, A, etc., be as before. Fix a compact subset L of $\mathbf{R}^n \backslash A$ which is of (Hausdorff) dimension $n - d - 1$. For instance, L might be a (standard round) sphere of dimension $n - d - 1$, or a polyhedron of dimension $n - d - 1$. Let us say that a compact subset E of $\mathbf{R}^n \backslash L$ is *linked* around L if there does not exist a way to continuously deform E into a point inside $\mathbf{R}^n \backslash L$. In other words, there should not exist a continuous one-parameter family of mappings f_t, $0 \leq t \leq 1$, from E into $\mathbf{R}^n \backslash L$ such that f_0 is the identity and $f_1(E)$ consists of only a single point.

Suppose now that E is contained in A. If E is linked around L, then E must satisfy (12.47). Otherwise, Proposition 12.46 would say that E could be homotoped inside A to a subset of a $(d-1)$-dimensional polyhedron, and this would imply that E itself could be contracted to a point in the complement of L (and therefore not be linked around L), by general-position results. (Compare with [**19, 18, 28**].)

One can also look at "linking" in more "dual" terms, analogous to the use of winding numbers in one complex variable. Here is a simple formulation. Imagine that we have a continuous mapping $\tau : A \to \mathbf{S}^d$, where $\mathbf{S}^d$ is the standard unit sphere in $\mathbf{R}^{d+1}$. We want to use τ to test the topological nondegeneracy of a compact subset E, by asking whether the restriction of τ to E is homotopically trivial, i.e., whether it can be continuously deformed to a constant mapping through maps from E into $\mathbf{S}^d$.

If the restriction of τ to E is homotopically nontrivial, then the same must be true for any image $f(E)$ of E under a mapping $f : E \to A$ which is homotopic to the identity through mappings from E into A. This is an easy exercise (which we also encountered in the previous section, in the proof of Lemma 12.57). On the other hand, any mapping from a $(d-1)$-dimensional polyhedron into $\mathbf{S}^d$ is homotopically trivial, as is well known (and not hard to prove). From this it follows that E has to satisfy (12.47) when the restriction of τ to E is homotopically nontrivial, because of Proposition 12.46.

If τ is Lipschitz, then one can use it to get an estimate like (12.47) directly, as follows. The restriction of τ to a set F is automatically homotopically trivial whenever $\tau(F)$ does not contain all of $\mathbf{S}^d$. This will happen as soon as the d-dimensional Hausdorff measure of F is small enough, i.e., when

$$(12.83) \qquad C_\tau^d H^d(F) < H^d(\mathbf{S}^d),$$

where C_τ is the Lipschitz constant for τ. Thus, if the restriction of τ to E is homotopically nontrivial, then

$$(12.84) \qquad H^d(E) \geq C_\tau^{-d} H^d(\mathbf{S}^d),$$

and in fact

$$(12.85) \qquad H^d(f(E)) \geq C_\tau^{-d} H^d(\mathbf{S}^d)$$

for every continuous mapping $f : E \to A$ which is homotopic to the identity through mappings from E into A. This last uses the observation mentioned above (in the paragraph preceding this one), that the restriction of τ to $f(E)$ is necessarily homotopically nontrivial if the restriction of τ to E is homotopically nontrivial, because of the homotopy condition on f.

In practice, the use of mappings to spheres as a test for topological nondegeneracy may proceed in a somewhat backwards manner compared to the description above. Instead of starting with the larger set A, one might begin with a compact set E and a homotopically nontrivial mapping $\tau : E \to \mathbf{S}^d$, and then try to choose $A \supseteq E$ and an extension of τ to A from the data of E and $\tau : E \to \mathbf{S}^d$. Part of the point would be to do this in such a way that A encompasses a neighborhood of E of definite size. Keep in mind that the extension of τ to A is supposed to have values in $\mathbf{S}^d$ too; if τ were allowed to take values in $\mathbf{R}^{d+1}$, for instance, then one could extend τ from E to all of $\mathbf{R}^n$ no matter the choice of E and τ. Such an extension would not be good for testing topological nondegeneracy, however.

It is easy to see that one can always extend τ as a mapping into $\mathbf{S}^d$ to some neighborhood of E, but to get a neighborhood of definite size one needs more information in general. If τ is actually Lipschitz on E, or has controlled modulus of continuity, then there is an extension of τ to a neighborhood of definite size, with bounds in terms of the modulus of continuity for τ. This is not hard to check.

Alternatively, there are some natural topological conditions on E which can control the size of the neighborhood for which one has an extension. For instance, if E admits a retraction from a larger set G onto itself, then τ automatically inherits an extension to G. Such a retraction can arise from bounds on the local contractability properties of E.

See [15, 23, 24, 26] and the references therein for further information related to some of these (topological) topics.

In [25], a particular question ("Conjecture 2") was formulated about rectifiability properties of sets (in general codimensions) under conditions of upper bounds for d-dimensional Hausdorff measure and lower bounds for d-dimensional topology, a question which is resolved in the affirmative by the present work (through Theorems 12.122 and 12.125 in Section 12.5, or (which is really the same thing) Theorem 0.10 in the introduction). The lower bounds for topology were much stronger than needed, and much stronger than the ones mentioned above. There it was assumed that E admit a homeomorphic parameterization by a d-dimensional ball, with bounds on the modulus of continuity for the homeomorphism as well as its inverse. Instead of using d-dimensional balls, it is easier to think of E as being homeomorphic to $\mathbf{S}^d$, and one can reduce to this case through a "doubling" construction. (That is, make two copies of E glued along the boundary. To avoid any danger of intersection of the two copies, one can do this inside of $\mathbf{R}^{n+1}$ instead of $\mathbf{R}^n$, with one copy in the hyperplane $x_{n+1} = 0$, and the other contained in the half-space $x_{n+1} > 0$ except along the boundary. For instance, one can simply define the x_{n+1} component to be the distance to the boundary of the disk.) If E is homeomorphic to $\mathbf{S}^d$, then one really only needs a bound for the modulus of continuity of the mapping from E to $\mathbf{S}^d$, as above, and only that this mapping be homotopically nontrivial, rather than a homeomorphism.

In any case, the formulation of the question in [25] was never intended to be definitive. It was purposely given relatively strong hypotheses and weak conclusions in order to serve as a test for a basic issue, i.e., so that it would not fail without a very good reason. As it is, the answer to this test question is positive, and the present methods provide a treatment of it which is more robust than the original formulation. In particular, the stability condition (12.47) is simpler and more robust than the one mentioned in [25] (and as indicated by the argument outlined above).

12.4. Polyhedral approximations and minimizers

Let A, j_0, d, etc., be as before.

LEMMA 12.86. *Let E be a compact subset of A with $H^d(E) < \infty$. Fix an integer $j \geq j_0$. Then there is a Lipschitz mapping $\zeta : E \to A$ with the following properties: (a) $\zeta(E \cap Q) \subseteq Q$ for all $Q \in \Delta_j$; (b) $\zeta(E) \subseteq \mathcal{S}_{j,d}$; (c) if T is one of the d-dimensional cubes of size 2^{-j} that make up $\mathcal{S}_{j,d}$, then either $\zeta(E)$ contains all of T, or $\zeta(E) \cap T \subseteq \partial T$; (d) $H^d(\zeta(E)) \leq C\, H^d(E)$, where the constant C depends* only on the dimension and not on A, j, or E; and (e) ζ is homotopic to the identity through mappings from E into A.

The existence of ζ can be derived from Lemma 11.14 in practically the same manner as for Lemma 12.49. Specifically, Lemma 11.14 (with this choice of E and j) provides a mapping f which satisfies (a), (b), and (d) in Lemma 12.86, but which need not satisfy (c). To get (c), one takes the d-dimensional cubes T in $\mathcal{S}_{j,d}$ such that $f(E)$ intersects but does not contain the interior of T, and one uses radial projections to clear out $f(E)$ from the interior of T. This is exactly like what happened in Lemma 12.49, except that now there may be some d-dimensional cubes T such that $f(E)$ contains all of T, and one leaves these alone.

Thus one can get a mapping ζ on E that satisfies (a) – (d) above. The fact that ζ takes values in A and the homotopy condition (e) follow from (a), just as in the remarks following the proof of Lemma 12.49. (See (12.56).) This completes the proof of Lemma 12.86.

REMARK 12.87. One could strengthen condition (c) in Lemma 12.86, so that $\zeta(E)$ is exactly a finite union of cubes which are faces of elements of Δ_j, cubes of dimension ranging between 0 and d. This can be obtained through exactly the same procedure of using radial projections to "clear out" compact subsets of a given cube R that do not contain R completely. In other words, instead of doing this only for the d-dimensional cubes, as above, one keeps going, to cubes of dimension $d-1$, then $d-2$, etc.

DEFINITION 12.88. *Let F be a compact subset of A. We shall call F j-simple if it is compact and if there is a finite sequence of d-dimensional cubes $T_1, T_2, \ldots, T_k$ in $\mathcal{S}_{j,d}$ (i.e., d-dimensional faces of cubes in Δ_j) such that F is the union of the T_i's together with a subset of $\mathcal{S}_{j,d-1}$.*

Lemma 12.86 provides a way to approximate a given set E by a set F which is j-simple. In this section we shall emphasize j-simple sets and other sets which are approximately polyhedral, and then return to general sets in Section 12.5.

Sets which are approximately polyhedral are nice for making it easier to obtain quasiminimizers with controlled properties, by taking minima of suitable functionals. Fix a compact j-simple set $F \subseteq A \cap \mathcal{S}_{j,d}$ and a number $M > 1$, and define $J_M(S)$ for closed sets S in $\mathbf{R}^n$ by

$$(12.89) \qquad J_M(S) = H^d(S \cap F) + M\, H^d(S \setminus F).$$

This is like the usual measurement of d-dimensional volume, except that we penalize the parts of S which lie outside of F.

This is the type of functional that we shall use, but we have to specify a class of competitors to go with it. Let $\mathcal{F}_j$ denote the collection of compact subsets X of A such that $X \subseteq \mathcal{S}_{j,d}$ and X is the image of F under a mapping $g : F \to A$ which is homotopic to the identity through mappings from F to A. Set

$$(12.90) \qquad \mathcal{F}_j^0 = \{X \in \mathcal{F}_j : X \text{ is } j\text{-simple}\}.$$

Note that $F \in \mathcal{F}_j^0$ by definition.

LEMMA 12.91. *For each $X \in \mathcal{F}_j$ there is an $X' \in \mathcal{F}_j^0$ such that $J_M(X') \leq J_M(X)$ for all M.*

This is like Lemma 12.86, but simpler. Let $X \in \mathcal{F}_j$ be given, and let T be one of the d-dimensional cubes that make up $\mathcal{S}_{j,d}$. If X contains T, or if X does not intersect the interior of T, then we leave T alone. Otherwise we push $X \cap T$ into ∂T using a radial projection onto ∂T, just as before. We do this for all T's, and the actions for the different T's do not conflict with each other, because the radial projections are all equal to the identity on the boundaries. (Keep in mind that $X \subseteq \mathcal{S}_{j,d}$ by assumption, so that one does not have to worry about other pieces of X.) This construction produces a new set $X' \subseteq A$ and a mapping ξ from X onto X', given by the radial projections on the parts of X which are being modified (and by the identity mapping on the portions of X which are not modified).

By construction, X' consists of the d-dimensional cubes $T_1, T_2, \ldots, T_k$ which are wholly contained in X together with a subset of $\mathcal{S}_{j,d-1}$. Thus X' is j-simple, and $J_M(X') \leq J_M(X)$ holds automatically for all M, since the part of X' contained in $\mathcal{S}_{j,d-1}$ has H^d-measure 0. That is, $H^d(X' \backslash X) = 0$.

To show that $X' \in \mathcal{F}_j^0$, the remaining point is that there is a mapping g' from F onto X' which is homotopic to the identity through mappings from F into A. There is such a mapping g from F onto X, since $X \in \mathcal{F}_j$ by assumption. For g' we simply take $\xi \circ g$. To get the homotopy condition it is enough to observe that $\xi : X \to A$ is homotopic to the identity through mappings from X into A. This is true for the same reasons as before, i.e., the trivial homotopy

$$(12.92) \qquad \Xi(x,s) = s\,\xi(x) + (1-s)\,x, \qquad 0 \leq s \leq 1,$$

works. More precisely, if $x \in A$ and Q is a cube in Δ_j such that $x \in Q$ and $Q \subseteq A$, then $\Xi(x,s)$ lies in Q for all $s \in [0,1]$, since x and $\xi(x)$ both do, by construction. Therefore $\Xi(x,s)$ lies in A for all $x \in X$ and $s \in [0,1]$, which is what we wanted. This completes the proof of Lemma 12.91.

REMARK 12.93. One can choose the set X' in Lemma 12.91 so that it is even a finite union of cubes of dimension $\leq d$ which are faces of cubes in Δ_j (and a finite polyhedron in particular). This is analogous to Remark 12.87.

LEMMA 12.94. *For each M there is an $X \in \mathcal{F}_j^0$ which minimizes J_M over $\mathcal{F}_j^0$.*

Indeed, since the elements of $\mathcal{F}_j^0$ are j-simple subsets of A, there are only finitely many of them modulo sets of H^d-measure 0 (i.e., modulo subsets of $\mathcal{S}_{j,d-1}$). The value of J_M is not changed by the addition or removal of sets of H^d-measure 0, and so the minimum exists because there are really only finitely many competing values for J_M. (This is the main reason for working with polyhedral approximations, to get the existence of minimizers automatically.)

Notice also that $\mathcal{F}_j^0$ is nonempty, because it contains our original set F as an element.

COROLLARY 12.95. *For each M there is an $X \in \mathcal{F}_j$ which minimizes J_M over $\mathcal{F}_j$.*

That is, a minimizer for $\mathcal{F}_j^0$ is also a minimizer for $\mathcal{F}_j$, because of Lemma 12.91.

There are now two main points to check, that minimizers for J_M in $\mathcal{F}_j$ are almost contained in our original set F, and that they are quasiminimizers with uniform bounds (and hence satisfy our regularity results with uniform bounds).

LEMMA 12.96. *If $X \in \mathcal{F}_j$ minimizes J_M, then*

$$(12.97) \qquad H^d(X \backslash F) \leq M^{-1} H^d(F).$$

Because F itself lies in $\mathcal{F}_j$ and X is a minimizer, we have that

(12.98) $$J_M(X) \leq J_M(F).$$

The inequality (12.97) follows from this and the definition (12.89) of J_M. (In particular, notice that $J_M(F) = H^d(F)$.)

In general Lemma 12.96 could be trivial, in the sense that the minimizer X has H^d-measure 0. This is exactly what will happen if F is not "stable" in the sense of (12.47). If F does satisfy (12.47), then we know that

(12.99) $$H^d(X) \geq \theta,$$

where θ is as in (12.47) and depends only on n and j_0. In particular, θ does not depend on F, j, or M, and (12.97) contains significant information (when M is sufficiently large).

We can do a bit better than this, using Proposition 12.61. Specifically,

(12.100) $$H^d_{con}(X) \geq \theta'$$

when F satisfies (12.47). Here $H^d_{con}(X)$ denotes the d-dimensional Hausdorff content of X, as in (12.58), and θ' is as in Proposition 12.61. Again, θ' depends only on n and j_0, and not on F, j, or M. (Strictly speaking, we are also using Lemma 12.57 in the derivation of (12.100), in addition to Proposition 12.61. This makes the bridge between stability for F and stability for X.)

We also know that the minimizer X should be "close" to F topologically, in the sense that it is the image of F under a mapping which is homotopic to the identity through maps from F into A (by definition of $\mathcal{F}_j$). In particular, natural topological nondegeneracy properties will carry over from F to X, as in Section 12.3. We shall discuss related themes in Remark 12.171 at the end of Section 12.5.

Note that if F arose through the application of Lemma 12.86 to a set $E \subseteq A$ which satisfies the stability condition (12.47), then F does too, because of Lemma 12.57 and part (e) of Lemma 12.86.

Recall that U is the open set defined at the beginning of Section 12.1, and that j_0 is also as in Section 12.1.

PROPOSITION 12.101. *Assume that $j \geq j_0$. If $X \in \mathcal{F}_j$ minimizes J_M over $\mathcal{F}_j$, then X is a $(U, k, +\infty)$-quasiminimizer (in the sense of Definition 1.9), where k depends only on n and M. In fact, if $\phi : X \to U$ is any continuous mapping which is homotopic to the identity through mappings from X into U, then (1.8) holds, i.e.,*

(12.102) $$H^d(X \cap W) \leq k\, H^d(\phi(X \cap W)),$$

where $W = \{x \in X : \phi(x) \neq x\}$.

It is very important here that k does not depend on X, F, or j. In the applications one can start with a compact set E in A, approximate it with a j-simple set F as in Lemma 12.86, and then obtain minimizers $X \in \mathcal{F}_j$ for J_M which also approximate E in a certain sense, as above. Proposition 12.101 says that these X's are quasiminimizers in U with constants that do not depend on E or on the resolution of the approximation of E by F (which is governed by j, as in Lemma 12.86). This permits us to obtain bounds on the geometry of the X's, independently of j, and to have some control in the limit as $j \to \infty$. We shall discuss this further in Section 12.5.

To prove Proposition 12.101, let us first show that (12.102) holds when $\phi : X \to U$ is homotopic to the identity through mappings from X into U and satisfies

(12.103) $$\phi(X) \subseteq A \cap S_{j,d}.$$

In this case ϕ is actually homotopic to the identity through mappings from X into A, since there is a retraction from U onto A, as in Lemma 12.2. (That is, a homotopy to the identity through mappings from X into U can be converted into a homotopy through mappings from X into A simply by composing with the retraction.) From here it follows that $\phi(X)$ lies in $\mathcal{F}_j$, i.e., $\phi(X)$ is the image of a continuous mapping from F into A which is homotopic to the identity through mappings from F into A. This is because X is the image of F by such a mapping, since $X \in \mathcal{F}_j$, and because of the homotopy condition for ϕ.

The minimality of X for J_M on $\mathcal{F}_j$ yields

(12.104) $$J_M(X) \leq J_M(\phi(X)).$$

Let W be as in the statement of Proposition 12.101. We can rewrite (12.104) as

(12.105) $$\begin{aligned} J_M(X\backslash W) + J_M(X \cap W) &= J_M(X) \\ &\leq J_M(\phi(X)) \\ &\leq J_M(\phi(X \cap W)) + J_M(\phi(X\backslash W)) \\ &= J_M(\phi(X \cap W)) + J_M(X\backslash W). \end{aligned}$$

The last equality uses the definition of W, i.e., $\phi(X\backslash W) = X\backslash W$. (Notice that the second inequality in (12.105) need not be an equality, because ϕ may not be injective and we are not counting H^d-measure with multiplicities.) Subtracting $J_M(X\backslash W)$ from both sides we get that

(12.106) $$J_M(X \cap W) \leq J_M(\phi(X \cap W)).$$

Since

(12.107) $$H^d \leq J_M \leq M H^d,$$

by the definition (12.89) of J_M, we conclude that

(12.108) $$H^d(X \cap W) \leq M H^d(\phi(X \cap W)).$$

This is the same as (12.102) with $k = M$.

The next step is to prove that (12.102) holds with

(12.109) $$\phi(X) \subseteq A$$

instead of (12.103), and with perhaps a larger value of k. For this one argues exactly as in Proposition 11.13. More precisely, let W be as above, and set $E = \overline{\phi(X \cap W)}$. As in the proof of Proposition 11.13, we apply Lemma 11.14 with these choices of j and E to get a mapping $f : \mathbf{R}^n \to \mathbf{R}^n$, and we set $\widetilde{\phi} = f \circ \phi$. Note that f maps A into A, because of (11.18) and the fact that A is a union of elements of Δ_j (since $j \geq j_0$ and A is a union of elements of Δ_{j_0}). This implies that $\widetilde{\phi}$ maps X into A, and in fact it maps X into $A \cap S_{j,d}$. Indeed, if $x \in X$, then either $x \in W$, in which case $\phi(x) \in E$ and $\widetilde{\phi}(x) = f(\phi(x))$ lies in $S_{j,d}$ by (11.17), or $x \in X\backslash W$, so that $\phi(x) = x$. In this event we use the fact that $X \subseteq S_{j,d}$, since $X \in \mathcal{F}_j$ by assumption, so that $x \in S_{j,d}$ and $\widetilde{\phi}(x) = f(x) = x$, by (11.16). Thus $\widetilde{\phi}(x)$ lies in $S_{j,d}$ in either circumstance, and $\widetilde{\phi}$ maps X into $A \cap S_{j,d}$.

12.4. POLYHEDRAL APPROXIMATIONS AND MINIMIZERS

We also have that $\widetilde{\phi}$ is homotopic to the identity through maps from X into A. To see this, it is enough to check that f is homotopic to the identity as a mapping from A into A, since ϕ is homotopic to the identity through maps into A by assumption. Consider the "obvious" homotopy

(12.110) $$tf(x) + (1-t)x, \quad 0 \le t \le 1.$$

This takes values in A for all $x \in A$ and $t \in [0,1]$. Indeed, given $x \in A$, let $Q \in \Delta_j$ be a cube such that $x \in Q$ and $Q \subseteq A$. Then $f(x)$ lies in Q, by (11.18), and hence (12.110) does too, since Q is convex. Thus f is homotopic to the identity through mappings from A into A, and $\widetilde{\phi}$ is also homotopic to the identity through maps into A.

In short, $\widetilde{\phi}$ satisfies the requirements of the first case, so that the analogue of (12.102) holds for $\widetilde{\phi}$ with $k = M$. Exactly as in the proof of Proposition 11.13, one can derive (12.102) for ϕ from this estimate for $\widetilde{\phi}$, with a choice of k which is equal to the product of M with a constant that depends only on the dimension. This corresponds to the part of the earlier argument that began at (11.21). (The portion before (11.21) was concerned with showing that $\widetilde{\phi}$ is an admissible deformation, which is simpler and involves fewer constraints in the present situation.)

It remains to show that (12.102) holds for any ϕ as in Proposition 12.101. This is analogous to Lemma 11.5, but it is somewhat simpler than that. Let $g : Y \to A$ be a Lipschitz retraction as in Lemma 12.2, and set $\widehat{\phi} = g \circ \phi$. Then $\widehat{\phi}$ maps X into A, and it is homotopic to the identity through such mappings, since ϕ is homotopic to the identity through mappings from X into U, by assumption. Thus the previous step applies to $\widehat{\phi}$, to give the analogue of (12.102) with ϕ replaced by $\widehat{\phi}$. One can derive (12.102) for ϕ itself from this version, exactly as in the proof of Lemma 11.5, starting at (11.11). (The notation is a bit different now, in that g here corresponds to h there, and $\widehat{\phi}$ was called $\widetilde{\phi}$ before. The h and ϕ in Lemma 11.5 enjoyed more properties than they do here, but these were used to guarantee the admissibility of $\widetilde{\phi}$, and they are not needed in the present situation.)

This completes the proof of Proposition 12.101. Note that we could also have used Lemma 11.5 and Proposition 11.13 in their earlier formulations, but with the slightly weaker conclusion then that X is a (U, k, δ)-quasiminimizer for suitable choices of k and δ. This would still be enough for the following conclusion.

COROLLARY 12.111. *Same assumptions and notations as in Proposition 12.101. Let X^* be as in (1.12), i.e., the subset of X which is the support of H^d restricted to X. Then X^* is a d-dimensional Ahlfors-regular set which is uniformly rectifiable and has big pieces of Lipschitz graphs (see Chapter 2 for the definitions), with constants that depend only on n, j_0, M, and the diameter of A.*

Again, it is very important here that the constants do not depend on X, F, or j.

Corollary 12.111 is an immediate consequence of Proposition 12.101 and Theorem 2.11. The extra dependence on j_0 and the diameter of A comes from the fact that Theorem 2.11 only provides control on the behavior of X^* at the scale of balls which sit well inside U, while the definitions of Ahlfors regularity and the rest permit radii up to the size of the diameter of X^*.

To summarize a bit, if we start with a compact set $E \subseteq A$ which is stable in the sense of (12.47) and satisfies $H^d(E) < \infty$, then we can approximate E by a j-simple set F using Lemma 12.86, and then approximate F by a set X which minimizes J_M in $\mathcal{F}_j$, as in Corollary 12.95 and Lemma 12.96. This set X is a quasiminimizing subset of U, in the sense of Definition 1.9, and with uniform control on the constants. As a result, we are able to apply Theorem 2.11 to obtain information about the structure of X^*, in terms of Ahlfors regularity, uniform rectifiability, and big pieces of Lipschitz graphs.

12.5. General sets

Let A, j_0, etc., be as before, and let E be a compact subset of A which satisfies the stability condition (12.47) and $H^d(E) < \infty$. Fix M, normally pretty large. For each $j \geq j_0$, let $\zeta_j : E \to A$ be as in Lemma 12.86, and set $F_j = \zeta_j(E)$. Thus F_j is a compact j-simple subset of A that approximates E. Let $\mathcal{F}_j$ be as in Section 12.4 (defined just before (12.90)), but with F_j in place of F. Let X_j be a minimizer of J_M over $\mathcal{F}_j$, as in Corollary 12.95. Note that F_j and X_j satisfy (12.47), since E does. This follows from Lemma 12.57, using also part (e) of Lemma 12.86 and the definition of $\mathcal{F}_j$. In particular, we have that

(12.112) $$H^d(X_j) \geq \theta$$

for all $j \geq 0$. We also have that

(12.113) $$H^d_{con}(X_j) \geq \theta'$$

for all $j \geq 0$, because of Proposition 12.61 and Lemma 12.57. (The latter provides bridges between the stability of E, F_j, and X_j.) Here $H^d_{con}(Z)$ denotes the d-dimensional Hausdorff content of Z, as in (12.58), and θ' is as in Proposition 12.61, and depends only on j_0 and n.

As in (1.12), write X_j^* for the subset of X_j which is the support of H^d restricted to X_j. In our case, X_j^* is a finite union of d-dimensional cubes from $\mathcal{S}_{j,d}$. (That is, whenever X_j^* does not completely contain some d-dimensional face T, it can only contain a set of H^d-measure 0 in T. This is because of minimality, and we have seen this kind of point before, in Lemma 12.91 and Remark 11.26.) Each X_j^* is nonempty, because of (12.112).

We want to take a limit of a subsequence of the X_j^*'s in the Hausdorff metric. Recall that if H, K are nonempty compact subsets of $\mathbf{R}^n$, then the Hausdorff distance $D(H, K)$ between them is defined to be the smallest number $t \geq 0$ such that

(12.114) $$\mathrm{dist}(x, H) \leq t \quad \text{for every } x \in K$$

and (symmetrically) $\mathrm{dist}(y, K) \leq t$ for every $y \in H$. It is well-known that the set of all nonempty compact subsets of a fixed compact set forms a compact metric space using the Hausdorff distance. For our purposes, this means that there is an infinite sequence $\{j_k\}_{k=1}^\infty$ and a nonempty compact set $Z \subseteq A$ such that

(12.115) $$\lim_{k \to \infty} D(X_{j_k}^*, Z) = 0.$$

Note that we are using the X_j^*'s instead of the X_j's in (12.115). The remainders $X_j \setminus X_j^*$ have H^d-measure zero and are not really controlled by the minimization of J_M which lead to X_j, and in the Hausdorff limit they could fill up large spurious sets. Such pathologies do not occur for the X_j^*'s, because of their Ahlfors regularity (with uniform bounds). We shall discuss this further in a moment.

12.5. GENERAL SETS

PROPOSITION 12.116. *Under the assumptions above,*

(12.117) $$H^d_{con}(Z) \geq \theta',$$

where θ' is as in Proposition 12.61 (and depends only on j_0 and n). In particular,

(12.118) $$H^d(Z) \geq \theta',$$

because of (12.59).

The proof of this is quite standard, given the analogous lower bound (12.113) for the X_j's, but let us go through it in some detail for the sake of completeness. Suppose that $H^d_{con}(Z)$ was strictly less than θ'. Then there would be a sequence $\{Y_\ell\}_\ell$ of sets in $\mathbf{R}^n$ which cover Z and satisfy

(12.119) $$\sum_\ell (\operatorname{diam} Y_\ell)^d < \theta',$$

by the definition (12.58) of $H^d_{con}(Z)$. We may as well assume that the Y_ℓ's are all open, since otherwise we can make them slightly larger so that this is true, while increasing their diameters only by such small amounts that (12.119) continues to hold. (These increases in the diameters can be chosen to zero quickly as $\ell \to \infty$.)

Since $\bigcup_\ell Y_\ell$ is open, Z is compact, and $Z \subseteq \bigcup_\ell Y_\ell$, there is an $\epsilon > 0$ so that any point in $\mathbf{R}^n$ which lies within ϵ of Z also lies in $\bigcup_\ell Y_\ell$. In particular, compact sets which are sufficiently close to Z in the Hausdorff metric are contained in $\bigcup_\ell Y_\ell$. From the definition of Z we have that this must be true for some of the X_j^*'s.

When $X_j^* \subseteq \bigcup_\ell Y_\ell$, we automatically have that

(12.120) $$H^d_{con}(X_j^*) \leq \sum_\ell (\operatorname{diam} Y_\ell)^d.$$

This yields $H^d_{con}(X_j^*) < \theta'$, because of (12.119). We also get that

(12.121) $$H^d_{con}(X_j) < \theta'.$$

This uses $H^d(X_j \setminus X_j^*) = 0$ (as in (1.13)), and the subadditivity of Hausdorff content. (Remember too that Hausdorff measure 0 implies Hausdorff content 0, as in (12.59).)

We now have a contradiction, since we already know that $H^d_{con}(X_j) \geq \theta'$, by (12.113). This proves Proposition 12.116.

This argument does not work with Hausdorff measure in place of Hausdorff content. The conclusion is not true either, i.e., this kind of simple lower semi-continuity of Hausdorff measure with respect to Hausdorff convergence of compact sets does not hold in general. We shall say more about this a bit below, shortly after the statement of Theorem 12.122. This is part of the trade-off between Hausdorff measure and Hausdorff content (and as mentioned before, in Section 12.2); Hausdorff content is more stable in this way, but less local, and not additive (in general circumstances) in particular.

In some situations, one can say that Z has the same kind of topological nondegeneracy as our original set E has. This will be discussed further in Remark 12.171. When this happens, one has another approach to lower bounds for the size of Z (as in Proposition 12.116). Namely, lower bounds for the size of Z would come directly from the topology, as in Sections 12.2 and 12.3, and just as for E and the X_j's. This is as opposed to trying to derive lower bounds for the size of Z from the lower bounds for the X_j's (which themselves come from topological considerations), as above.

THEOREM 12.122. *Under the assumptions above, Z is an Ahlfors-regular set of dimension d which is uniformly rectifiable and has big pieces of Lipschitz graphs, all with bounds that depend only on n, j_0, M, and the diameter of A.*

We know already from Corollary 12.111 that the X_j^*'s enjoy the properties in Theorem 12.122 for j large enough, and with uniform bounds. The point here is simply to pass to the limit. This is rather straightforward, if slightly tedious and technical, and we shall only describe the key ingredients. A more thorough treatment of these matters can be found in [**11**], especially Chapter 8 and Section 9.3.

In general, Hausdorff measures can either go up or down in Hausdorff limits. Sequences of finite sets can converge to whole cubes, for instance, by taking more and more dense approximations. To have Hausdorff measures go down dramatically, consider the sequence of curves $\{\Gamma_j\}_{j=1}^\infty$ in the plane given by

$$(12.123) \qquad \Gamma_j = \{(x,y) \in \mathbf{R}^2 : 0 \leq x \leq 1,\, y = j^{-1} \sin j^2 x\}.$$

These curves converge in the Hausdorff sense to the segment that goes from $(0,0)$ to $(1,0)$. This segment has 1-dimensional Hausdorff measure equal to 1, while $H^1(\Gamma_j)$ is bounded from below by a multiple of j as $j \to \infty$. (This should be compared with the situation for Hausdorff content instead of Hausdorff measure, as in Proposition 12.116 and its proof.)

For the proof of Theorem 12.122, the main point is that the possibility of such pathologies is severely limited in the case of Ahlfors-regular sets. In particular, the Hausdorff limit of a sequence of Ahlfors-regular sets of dimension d is also Ahlfors regular of dimension d, at least if the regularity constants for the sets in the sequence remain bounded. One way to prove this is to reformulate the Ahlfors regularity of a set H in terms of upper and lower bounds for the number of balls of radius r needed to cover $H \cap B(x, R)$ when $x \in H$ and $0 < r \leq R \leq \operatorname{diam} H$. Specifically, this number should be bounded from above and below by constant multiples of $(R/r)^d$ when H is regular of dimension d, as one can check. This type of condition is easily seen to be preserved by Hausdorff limits, when one has uniform bounds for the constants involved.

For *subsets* of Ahlfors-regular sets one also has some control. Assume, for instance, that $\{K_\ell\}$ is a sequence of compact subsets of A which converge in the Hausdorff metric to a compact set K, and that the K_ℓ's are all contained in Ahlfors-regular sets of dimension d with bounded constants (like the X_j^*'s). Then

$$(12.124) \qquad H^d(K) \geq C^{-1} \limsup_{\ell \to \infty} H^d(K_\ell),$$

where C depends only on d and the regularity constants involved. This is not hard to prove, straight from the definitions. (See also Lemma 8.68 in Section 8.7 of [**11**].)

One can also look at (12.124) in the following terms. The d-dimensional Hausdorff measure of a set is always greater than or equal to the d-dimensional Hausdorff content, as in (12.59). For subsets of an Ahlfors-regular set of dimension d, Hausdorff measure and Hausdorff content in dimension d are approximately the same, to within bounded factors. This is not hard to see, just from the definitions, and it was mentioned earlier in Section 12.2, shortly after the definition (12.58) of Hausdorff content. Using these pieces of information, one can derive (12.124) from the general lower-semicontinuity for Hausdorff content (without the constant, as in the proof of Proposition 12.116).

Note that (12.124) can be used to give a lower bound for the Hausdorff measure of Z, as in (12.118). However, this lower bound is not as good as the one in (12.118), because of the constants involved. It would give a bound which depends on M, while (12.118) does not depend on M.

Using (12.124), one can derive the uniform rectifiability and BPLG properties of Z in Theorem 12.122 from the corresponding conditions for the $X_{j_k}^*$'s (and with uniform bounds). More precisely, these conditions are defined in terms of the existence of subsets of Z with substantial measure and controlled geometry, as in Definitions 2.3 and 2.7. Using (12.124), one can obtain these sets in Z as limits of their counterparts in the $X_{j_k}^*$'s. To check this carefully one also has to work with limits of sequences of Lipschitz and bilipschitz mappings, but this is not too difficult. In particular, there are uniform bounds for the Lipschitz and bilipschitz constants, which provide the equicontinuity needed for arguments of Arzela-Ascoli type. For all of these limits one is free to pass to subsequences as needed to obtain convergence.

This completes the sketch of the proof of Theorem 12.122. See [11] for more details.

THEOREM 12.125. *Same assumptions and notations as above. There is a constant C_1 so that*

(12.126) $$H^d(Z\backslash E) \leq C_1 M^{-1} H^d(E),$$

where C_1 depends only on n.

In other words, nearly all of Z is contained in E when M is large. It is essential here that C_1 be independent of M. To get this, we shall have to go back and look at the Ahlfors-regularity properties of the X_j^*'s more carefully.

Overall, the price for taking M large comes in the constants for Z provided by Theorem 12.122. These constants depend on the quasiminimizing parameter k for the X_j's in Proposition 12.101, and this in turn depends on M through (12.108). For the purposes of Theorem 12.125, however, we only have to worry about what happens away from E. This will permit us to avoid the "nonuniformity" of the functional J_M in (12.89), and to get local bounds on the X_j's (away from E) which do not depend on M.

To make this precise, let $\epsilon > 0$ be given, and set

(12.127) $$Z(\epsilon) = \{x \in Z : \text{dist}(x, E) \geq 3\,\epsilon\}.$$

If we can show that

(12.128) $$H^d(Z(\epsilon)) \leq C_1 M^{-1} H^d(E)$$

for all sufficiently small ϵ, with C_1 depending only on n (and not on ϵ in particular), then (12.126) will follow.

In fact we shall show that for all small $\eta > 0$ there is a covering of $Z(\epsilon)$ by at most

(12.129) $$C_2\, \eta^{-d}\, M^{-1} H^d(E)$$

balls of radius η, where C_2 depends only on n. If we can do that, then (12.128) will follow, by the definition (0.2) of Hausdorff measure. Note that we may restrict ourselves to η's which are as small as we like, and $\eta < \epsilon$ in particular. (In other words, η is allowed to depend on ϵ (and, of course, not the reverse).)

This formulation lends itself more easily to the definition of Z as the Hausdorff limit of the $X_{j_k}^*$'s. Set

(12.130) $$Z_j(\epsilon) = \{x \in X_j^* : \operatorname{dist}(x, E) \geq 2\epsilon\}.$$

We want to show that $Z_j(\epsilon)$ can be covered by at most

(12.131) $$C_3 \, \eta^{-d} \, M^{-1} H^d(E)$$

balls of radius η for all sufficiently large j, where C_3 does not depend on ϵ, η, M, or j. If we can prove this, then the corresponding statement for $Z(\epsilon)$ will follow. To see this, we observe first that

(12.132) $$D(Z, X_{j_k}^*) < \min(\epsilon, \eta) \qquad \text{when } j_k \text{ is sufficiently large,}$$

by Hausdorff convergence. Using this it is easy to check that

(12.133) $$\operatorname{dist}(x, Z_{j_k}(\epsilon)) < \eta \qquad \text{for all } x \in Z(\epsilon)$$

when j_k is sufficiently large. This implies that a covering of $Z_j(\epsilon)$ by N balls of radius η leads to a covering of $Z(\epsilon)$ by N balls of radius 2η when j_k is large enough. In other words, coverings for $Z_j(\epsilon)$ lead to coverings for $Z(\epsilon)$, with suitable estimates. Thus we are reduced to showing that $Z_j(\epsilon)$ can be covered by at most (12.131) balls of radius η for all sufficiently large j.

Next, let us check that points in $Z_j(\epsilon)$ never get too close to F_j when j is large enough. From part (a) of Lemma 12.86 we have that

(12.134) $$\operatorname{dist}(y, E) \leq \sqrt{n}\, 2^{-j} \qquad \text{for all } y \in F_j.$$

If J is large enough, then we obtain that

(12.135) $$\operatorname{dist}(y, E) < \epsilon \qquad \text{for all } y \in F_j$$

when $j \geq J$, so that

(12.136) $$Z_j(\epsilon) \subseteq \{x \in X_j^* : \operatorname{dist}(x, F_j) \geq \epsilon\}$$

when $j \geq J$. Thus we are now reduced to showing that

(12.137) $$\{x \in X_j^* : \operatorname{dist}(x, F_j) \geq \epsilon\}$$

can be covered by at most (12.131) balls of radius η.

On the other hand, we know from Lemma 12.96 that

(12.138) $$H^d(X_j^* \setminus F_j) \leq M^{-1} H^d(F_j).$$

We also have that $H^d(F_j) \leq C\, H^d(E)$, where C depends only on n, since F_j was obtained from E through Lemma 12.86. Thus

(12.139) $$H^d(X_j^* \setminus F_j) \leq C\, M^{-1} H^d(E)$$

for all j.

Suppose for the moment that we can show that

(12.140) $$H^d(X_j^* \cap B(x, \eta/2)) \geq C_4^{-1} \eta^d$$

for all $x \in X_j^*$ and j with $\operatorname{dist}(x, F_j) \geq \epsilon$, and for all sufficiently small η, where C_4 depends only on n, and not on ϵ, η, j, or M. How small η has to be will depend on n, j_0, and ϵ, but nothing else, and not on x in particular. If we can do this, then we can get the covering that we need for (12.137) from the estimate (12.139). This follows from a standard argument, which we now recall. Let $\{x_i\}_{i=1}^N$ be a maximal subset of (12.137) such that

(12.141) $$|x_i - x_\ell| \geq \eta \quad \text{when } i \neq \ell.$$

Then (12.137) is covered by the balls $B(x_i, \eta)$, $1 \leq i \leq N$, because otherwise $\{x_i\}_{i=1}^N$ would not be maximal; i.e., an omitted point could be added to the x_i's without disrupting (12.141), contradicting maximality. Thus we do have a covering, and we only need to get a bound for N. For this we observe that the balls $B(x_i, \eta/2)$, $1 \leq i \leq N$, are pairwise disjoint, by (12.141), and that each $B(x_i, \eta/2)$ is contained in the complement of F_j, at least if $\eta < \epsilon$, since the x_i's are contained in (12.137). Therefore

(12.142) $$\sum_{i=1}^N H^d(B(x_i, \eta/2) \cap X_j^*) = H^d\left(\bigcup_{i=1}^N B(x_i, \eta/2) \cap X_j^*\right)$$
$$\leq C M^{-1} H^d(E),$$

by (12.139). This shows that (12.140) implies a bound for N like (12.131), which is what we want.

We are left with proving (12.140). This can be viewed as a piece of an Ahlfors-regularity condition, which we can seek to derive from quasiminimality considerations, as in Theorem 2.11. The point is to have quasiminimality constants that do not depend on M, so that the Ahlfors-regularity constants do not depend on M either.

Fix a j, large, and a z in X_j^* such that

(12.143) $$\text{dist}(z, F_j) \geq \epsilon,$$

and set $B = B(z, 2\eta)$. We want to show that

(12.144) $$X_j \cap B \text{ is a } (B, k, +\infty)\text{-quasiminimizer for } H^d$$

when

(12.145) $$6\eta < \epsilon \quad \text{and} \quad 4\eta < \delta_1,$$

where δ_1 comes from Lemma 12.23 (with $\delta = +\infty$) and depends only on n and j_0. The constant k in (12.144) depends on n, but nothing else, and not on ϵ, η, j, M, or z in particular.

To prove (12.144), one argues in nearly the same manner as for Proposition 12.101, using the methods of Lemma 11.5 and Proposition 11.13. Unfortunately, we cannot apply Lemma 11.5 and Proposition 11.13 directly to B (in place of U), but we can still use the constructions that they provide, and that is what we shall do.

Let $\phi : \mathbf{R}^n \to \mathbf{R}^n$ be a mapping which satisfies (1.5), (1.6), and (1.7), but with U replaced with B. In the present circumstances, (1.5) is vacuous, since $\delta = +\infty$, but (1.6) implies that

(12.146) $$W_\phi \cup \phi(W_\phi) \subseteq B,$$

where $W_\phi = \{x \in \mathbf{R}^n : \phi(x) \neq x\}$. We want to show that (1.8) holds, i.e., that

(12.147) $$H^d(X_j \cap W_\phi) \leq k H^d(\phi(X_j \cap W_\phi)).$$

(We do not need to include the intersection with B here, because that is accounted for by W_ϕ.) We assume that

(12.148) $$X_j \cap W_\phi \neq \emptyset,$$

since otherwise (12.147) is trivial.

As in Lemma 11.5, we want to convert ϕ into a mapping $\widetilde{\phi} : \mathbf{R}^n \to \mathbf{R}^n$ such that

(12.149) $$\widetilde{\phi}(X_j) \subseteq A.$$

We define $\widetilde{\phi}$ in exactly the same way as in the proof of Lemma 11.5, namely by taking

(12.150) $$\widetilde{\phi} = h_L \circ \phi,$$

where L is the closure of $W_\phi \cup \phi(W_\phi)$ and h_L is as in the hypothesis of Lemma 11.5. In this case this means that h_L is provided by Lemma 12.23 (with $\delta = \infty$). Note that L intersects X_j, by (12.148), and therefore intersects A in particular, as required in Lemmas 11.5 and 12.23. Also,

(12.151) $$\operatorname{diam} L \leq 4\eta,$$

because of (12.146), and this gives $\operatorname{diam} L < \delta_1$ in Lemma 12.23, by (12.145).

From Lemma 12.23 we have that the Lipschitz norm of h_L is bounded by a constant that depends only on n, and not on η, ϵ, j, z, or anything like that. This will be important later on.

Let us check that

(12.152) $$|y - z| \leq 6\eta \quad \text{when } h_L(y) \neq y.$$

Indeed, if $h_L(y) \neq y$, then $y \in D$, as mentioned just after (12.27). The set D was defined just before (12.24), as the collection of points at distance $\leq \operatorname{diam} L$ from L. Thus $h_L(y) \neq y$ implies that $\operatorname{dist}(y, L) \leq \operatorname{diam} L$, so that

(12.153) $$\operatorname{dist}(y, L) \leq 4\eta,$$

because $\operatorname{diam} L \leq 4\eta$. Since we took L to be the closure of $W_\phi \cup \phi(W_\phi)$, we have that L is contained in $\overline{B} = \overline{B}(z, 2\eta)$, by (12.146). This proves (12.152), because of (12.153).

Next we want to convert $\widetilde{\phi}$ into a mapping $\widetilde{\phi}_1 : \mathbf{R}^n \to \mathbf{R}^n$ such that

(12.154) $$\widetilde{\phi}_1(X_j) \subseteq A \cap \mathcal{S}_{j,d}.$$

We do this exactly as in the proof of Proposition 11.13, by taking

(12.155) $$\widetilde{\phi}_1 = f \circ \widetilde{\phi},$$

where $f : \mathbf{R}^n \to \mathbf{R}^n$ is as in Lemma 11.14. For this application of Lemma 11.14 we use the present choice of j, and we take E to be the closure of $\widetilde{\phi}(X_j \cap W_{\widetilde{\phi}})$, with $W_{\widetilde{\phi}} = \{x \in \mathbf{R}^n : \widetilde{\phi}(x) \neq x\}$. Notice that

(12.156) $$\widetilde{\phi}(X_j) \subseteq \widetilde{\phi}(X_j \cap W_{\widetilde{\phi}}) \cup X_j,$$

since $\widetilde{\phi}(y) = y$ on the complement of $W_{\widetilde{\phi}}$. Thus

(12.157) $$\begin{aligned} \widetilde{\phi}_1(X_j) = f(\widetilde{\phi}(X_j)) &\subseteq f(\widetilde{\phi}(X_j \cap W_{\widetilde{\phi}})) \cup f(X_j) \\ &\subseteq f(\widetilde{\phi}(X_j \cap W_{\widetilde{\phi}})) \cup X_j \end{aligned}$$

The latter inclusion uses the fact that $X_j \subseteq \mathcal{S}_{j,d}$, so that f is equal to the identity on X_j, by (11.16). From (12.157) we obtain that $\widetilde{\phi}_1(X_j)$ is contained in $\mathcal{S}_{j,d}$, since $f(\widetilde{\phi}(X_j \cap W_{\widetilde{\phi}}))$ is, by (11.17). We also have that $\widetilde{\phi}_1(X_j)$ is contained in A, because $\widetilde{\phi}(X_j) \subseteq A$, as in (12.149), and because f maps A into A, by (11.18). (This last also uses the inequality $j \geq j_0$ to know that A is a union of cubes in Δ_j.) Altogether we conclude that (12.154) does hold.

We want to use the minimality of X_j in $\mathcal{F}_j$ for J_M to get that

$$(12.158) \qquad J_M(X_j) \leq J_M(\widetilde{\phi}_1(X_j)).$$

To do this, we need to know that the restriction of $\widetilde{\phi}_1$ to X_j is homotopic to the identity through mappings of X_j into A, so that $\widetilde{\phi}_1(X_j)$ is an acceptable competitor. This one can get from the original homotopy for ϕ from (1.7) exactly as in the proof of Proposition 12.101, using also the homotopies for h_L and f that one can get from Lemmas 12.23 and 11.14, and the retraction of U onto A from Lemma 12.2. (Alternatively, for j large and η small one can get the existence of a homotopy from Lemma 12.42 and the fact that $\widetilde{\phi}_1$ never moves points very far. The latter is made precise by (12.146), (11.18), and the proof of Lemma 12.23 (especially (12.32), (12.39), for instance).) Thus we have (12.158).

Set $W_{\widetilde{\phi}_1} = \{x \in \mathbf{R}^n : \widetilde{\phi}_1(x) \neq x\}$. A crucial point is to show that (12.158) implies

$$(12.159) \qquad H^d(X_j \cap W_{\widetilde{\phi}_1}) \leq H^d(\widetilde{\phi}_1(X_j \cap W_{\widetilde{\phi}_1})),$$

i.e., without M or anything like that. The counterpart of (12.159) in the proof of Proposition 12.101 was (12.108), and it was exactly the place where the dependence on M came about.

To establish (12.159), we begin by observing that (12.158) implies

$$(12.160) \qquad J_M(X_j \cap W_{\widetilde{\phi}_1}) \leq J_M(\widetilde{\phi}_1(X_j \cap W_{\widetilde{\phi}_1})).$$

This can be checked exactly as in (12.105)–(12.106).

Remember from (12.89) that J_M is defined by

$$(12.161) \qquad J_M(S) = H^d(S \cap F_j) + M\, H^d(S \backslash F_j),$$

and that M is always at least 1. Thus (12.160) yields

$$(12.162) \quad H^d(X_j \cap W_{\widetilde{\phi}_1} \cap F_j) + M\, H^d((X_j \cap W_{\widetilde{\phi}_1}) \backslash F_j)$$
$$= J_M(X_j \cap W_{\widetilde{\phi}_1}) \leq J_M(\widetilde{\phi}_1(X_j \cap W_{\widetilde{\phi}_1}))$$
$$\leq M\, H^d(\widetilde{\phi}_1(X_j \cap W_{\widetilde{\phi}_1})).$$

If we can show that

$$(12.163) \qquad X_j \cap W_{\widetilde{\phi}_1} \subseteq \mathbf{R}^n \backslash F_j,$$

then (12.159) will follow from (12.162).

Let x be any element of $X_j \cap W_{\widetilde{\phi}_1}$, and let us verify that

$$(12.164) \qquad |x - z| \leq 6\eta.$$

Since $x \in W_{\widetilde{\phi}_1}$, we have that $\widetilde{\phi}_1(x) \neq x$. One of the following three conditions must therefore be satisfied:

(12.165) $\quad\quad\quad\quad\quad \phi(x) \neq x;$

(12.166) $\quad\quad\quad\quad\quad \phi(x) = x$ but $h_L(x) \neq x;$

(12.167) $\quad\quad\quad\quad\quad \phi(x) = x$, $h_L(x) = x$, but $f(x) \neq x$.

(Remember that $\widetilde{\phi}_1 = f \circ h_L \circ \phi$, by construction.) If (12.165) holds, then x lies in $B = B(z, 2\eta)$, by (12.146), which gives (12.164). If instead (12.166) is true, then we get (12.164) from (12.152). The third alternative (12.167) cannot occur. Indeed, x lies in X_j, and hence in $\mathcal{S}_{j,d}$. This means that $f(x) = x$, by (11.16), in contradiction to (12.167).

This proves that (12.164) holds when $x \in X_j \cap W_{\widetilde{\phi}_1}$. On the other hand,

(12.168) $\quad\quad\quad\quad\quad 6\eta < \mathrm{dist}(z, F_j),$

by (12.143) and (12.145). Thus we get (12.163), and (12.159) follows.

It remains to show that (12.159) can be converted into (12.147). This follows exactly the same arguments as in Proposition 11.13 and Lemma 11.5. More precisely, the first step is to go from (12.159) to

(12.169) $\quad\quad\quad\quad H^d(X_j \cap W_{\widetilde{\phi}}) \leq C\, H^d(\widetilde{\phi}(X_j \cap W_{\widetilde{\phi}}))$

for a suitable constant C (which depends only on the dimension). This is essentially the same as in the proof of Proposition 11.13, starting with (11.21) and ending with (11.25). The second step is to go from (12.169) to

(12.170) $\quad\quad\quad\quad H^d(X_j \cap W_\phi) \leq C'\, H^d(\phi(X_j \cap W_\phi)).$

This is practically the same as in the proof of Lemma 11.5, starting at (11.11). As in Lemma 11.5, the constant C' can be computed from the preceding one C and a bound for the Lipschitz norm of h_L. Thus C' can be chosen so that it depends only on n, because of the bound for the Lipschitz norm of h_L from Lemma 12.23.

This completes the proof of (12.147). We conclude that (12.144) holds, with a constant k that depends only on n, under the assumption (12.145). Using this and Theorem 2.11 we get that (12.140) is true when η is small enough, j is large enough, and $x \in X_j^*$ satisfies $\mathrm{dist}(x, F_j) \geq \epsilon$. (Note that the relationship between the balls in (12.140) and (12.144) – i.e., the ball B in (12.144) can be taken to be the quadruple of the one in (12.140) – is compatible with the condition (2.12) in Theorem 2.11.) This is sufficient to get a covering of (12.137) by at most (12.131) balls of radius η, as discussed before. Theorem 12.125 follows, by our earlier arguments.

Note that (12.144) leads to somewhat better control on the local geometry of Z away from E than is provided by Theorem 12.122, better in the sense that the constants involved do not depend on M.

REMARK 12.171. Theorem 12.125 and Proposition 12.116 imply that our limiting set Z contains a piece of the original set E of definite size. In practice one can say more than that, with the rough idea being that Z carries the same kind of nontrivial d-dimensional topology that E does. In the codimension-1 case (where $d = n - 1$), for instance, Z will separate the same pairs of points in the complement of A as E does. This was one of the starting points in [9].

Let us look at this in a more careful way. For the X_j's as above the relationship with E is more explicit, and each X_j is the image of E under a mapping $\xi_j : E \to A$

which is homotopic to the identity through mappings from E into A. This is because the F_j's are images of E under such mappings, by construction (see the beginning of the section and Lemma 12.86), and because each X_j is the image of F_j by a mapping which is homotopic to the identity through mappings from F_j into A, since $X_j \in \mathcal{F}_j$. (Remember that $\mathcal{F}_j$ is defined just before (12.90), and that we are now taking $F = F_j$.)

One could strengthen this somewhat, by adjusting the definitions so that the X_j's are images of E under maps from A to A that are homotopic to the identity through mappings from A to itself (instead of just going from E to A). For this one should work with mappings from A to A throughout, i.e., in Proposition 12.46 and Lemma 12.86, and in the definition of $\mathcal{F}_j$. The proofs of the new versions of the latter would require a few extra steps, but this would not be very difficult (and similar to parts of the proof of Proposition 3.1). The same overall analysis could be made as before.

Now let us think about the set Z. In the transition from the X_j's to Z, there are two main changes which take place. Namely, the replacement of X_j with X_j^*, and then the passage to the limit. The substitution of X_j^* for X_j is not very significant for d-dimensional topology, because the part of X_j which is pruned to give X_j^* is of lower topological dimension. Indeed, $X_j \backslash X_j^*$ has H^d-measure 0; if one is (slightly) careful to choose X_j so that it lies in $\mathcal{F}_j^0$, then $X_j \backslash X_j^*$ is a subset of $\mathcal{S}_{j,d-1}$. The passage to the limit is also compatible with many conditions of topological nondegeneracy. If degeneracy of Z is somehow measured by the existence of a continuous mapping from Z into another space, for instance, then the existence of such a mapping on Z typically leads to an extension on a neighborhood of Z, and hence to the same kind of degeneracy for sets which are close to Z in the Hausdorff topology.

For example, let us consider the stability condition (12.47). If the X_j^*'s satisfy (12.47), then it is not hard to show that Z would too. To see this, assume that Z fails to satisfy (12.47), so that there is a continuous mapping from Z into $A \cap \mathcal{S}_{\ell,d-1}$ for some $\ell \geq j_0$ which is homotopic to the identity through maps from Z into A, as in Proposition 12.46. It is not hard to show that there is then an extension of this mapping which takes a (small) neighborhood V of Z into $A \cap \mathcal{S}_{\ell,d-1}$ and which is homotopic to the identity through mappings from V into A. Since Z is the Hausdorff limit of a subsequence of $\{X_j^*\}$, we conclude that $X_j^* \subseteq V$ for infinitely many values of j, and this contradicts the assumption that the X_j^*'s satisfy (12.47).

Unfortunately, in general the validity of (12.47) for the X_j's might not imply the corresponding assertion for the X_j^*'s. This is a fairly minor technicality, but the class of mappings from an X_j^* into A which are homotopic to the identity could be considerably larger than the class of mappings from X_j into A which are homotopic to the identity. This would not be a problem if the homotopy groups of A in dimensions $< d$ were trivial. One can avoid this problem altogether by working with mappings from A to A throughout, as discussed above, rather than from subsets of A into A. That is, the modified version of (12.47) would not be disturbed by the substitution of X_j^* for X_j. This is not too hard to check. (For instance, one can use Lipschitz mappings here, rather than general continuous ones, through approximation if need be. With Lipschitz mappings, there is no danger that $X_j \backslash X_j^*$ is accidentally deformed into a set of positive H^d-measure.)

For other types of topological nondegeneracy conditions – in terms of linking or homotopically-nontrivial mappings into $\mathbf{S}^d$, for example, as in Section 12.3 – the substitution of X_j^* for X_j is even simpler, and quite standard (and can be made more directly). For this it is helpful to think of X_j as being j-simple, so that $X_j \backslash X_j^*$ is contained in $\mathcal{S}_{j,d-1}$.

Note that for nondegeneracy conditions of linking around a polyhedron, or homotopic-nontriviality of a mapping into $\mathbf{S}^d$, as in Section 12.3, there is no problem with passing to Hausdorff limits. That is, if the limit Z were degenerate, then this could be expressed in terms of the existence of a continuous mapping on Z with certain properties. This mapping could be extended to a neighborhood of the set in question, and this would lead to analogous degeneracy for sets which are close to Z in the Hausdorff sense (assuming that Z is compact).

For mappings *into* moving sets, the effects of Hausdorff convergence is quite different. In particular, although the X_j's are continuous (or even Lipschitz) images of E in the context of this section, the same may not be true for Hausdorff limits of the X_j's. It would be true if one had uniform bounds on the moduli of continuity of these mappings, but in general that need not be the case. This makes it harder to say that the Hausdorff limits really contain only the "same" topological information as the original sets, even if they often do satisfy much the same nondegeneracy properties, as above.

One could also work with currents and homology classes, rather than sets and homotopies, in order to control what happens topologically under limits. For that matter, by working with currents, one would be able to obtain existence of minimizers more directly, as in [13], rather than restricting to essentially finite classes of polyhedral sets first, as we have done here.

Bibliography

[1] F. Almgren, *Existence and regularity almost everywhere of solutions to elliptic variational problems among surfaces of varying topological type and singularity structure*, Annals of Mathematics (2) **87** (1968), 321–391.

[2] F. Almgren, *Existence and regularity almost everywhere of solutions to elliptic variational problems with constraints*, Memoirs of the American Mathematical Society **4** (1976), no. 165.

[3] *Seminar on Minimal Submanifolds*, edited by E. Bombieri, Annals of Mathematics Studies **103** (1983), Princeton University Press.

[4] G. David, *Morceaux de graphes lipschitziens et intégrales singulières sur un surface*, Revista Matemática Iberoamericana **4** (1988), 73-114.

[5] G. David, *Wavelets and Singular Integrals on Curves and Surfaces*, Lecture Notes in Mathematics **1465** (1991), Springer-Verlag.

[6] G. David and S. Semmes, *Singular Integrals and Rectifiable Sets in $\mathbf{R}^n$: au-delà des graphes lipschitziens*, Astérisque **193**, Société Mathématique de France, 1991.

[7] G. David and S. Semmes, *Quantitative rectifiability and Lipschitz mappings*, Transactions of the Amererican Mathematical Society **337** (1993), 855-889.

[8] G. David and S. Semmes, *Analysis of and on Uniformly Rectifiable Sets*, Mathematical Surveys and Monographs **38**, American Mathematical Society, 1993.

[9] G. David and S. Semmes, *Quasiminimal surfaces of codimension 1 and John domains*, Pacific Journal of Mathematics **183** (1998), 213-277.

[10] G. David and S. Semmes, *Surfaces quasiminimales de codimension 1: un morceau de démonstration*, Journées "Équations aux Dérivées Partielles" (Saint-Jean-de-Monts, 1996), Exp. No. IX, École Polytechnique, Palaiseau, 1996.

[11] G. David and S. Semmes, *Fractured Fractals and Broken Dreams: Self-Similar Geometry through Metric and Measure*, Oxford Lecture Series in Mathematics and its Applications **7**, Oxford University Press, 1997.

[12] K. Falconer, *The Geometry of Fractal Sets*, Cambridge University Press, 1984.

[13] H. Federer, *Geometric Measure Theory*, Springer-Verlag, 1969.

[14] E. Giusti, *Minimal Surfaces and Functions of Bounded Variation*, Birkhäuser, 1984.

[15] R. Greene and P. Petersen V, *Little topology, big volume*, Duke Mathematical Journal **67** (1992), 273-290.

[16] W. Hurewicz and H. Wallman, *Dimension Theory*, Princeton University Press, 1941.

[17] P. Jones, N. Katz, and A. Vargas, *Checkerboards, Lipschitz functions, and uniform rectifiability*, Revista Matemática Iberoamericana **13** (1997), 189-210.

[18] J. Luukkainen and J. Väisälä, *Elements of Lipschitz topology*, Annales Academiæ Scientiarum Fennicæ Mathematica Ser. A I Math. **3** (1977), 85-122.

[19] O. Martio, S. Rickman, and J. Väisälä, *Topological and metric properties of quasiregular mappings*, Annales Academiæ Scientiarum Fennicæ Mathematica Ser. A I Math. **488** (1971), 1-31.

[20] P. Mattila, *Geometry of Sets and Measures in Euclidean Spaces*, Cambridge University Press, 1995.

[21] F. Morgan, *Geometric Measure Theory: A Beginner's Guide*, Academic Press, 1988.

[22] J.-M. Morel and S. Solimini, *Variational Methods in Image Segmentation*, Birkhäuser, 1995.

[23] P. Petersen V, *A finiteness theorem for metric spaces*, Journal of Differential Geometry **31** (1990), 387-395.

[24] P. Petersen V, *Gromov-Hausdorff convergence of metric spaces*, Proceedings of Symposia in Pure Mathematics **54**, Part 3, (1993), 489-504, American Mathematical Society.

[25] S. Semmes, *Finding structure in sets with little smoothness*, Proceedings of the International Congress of Mathematicians (Zürich, 1994), Birkhäuser, 1995, p875-885.

[26] S. Semmes, *Finding curves on general spaces through quantitative topology, with applications to Sobolev and Poincaré inequalities*, Selecta Mathematica (N.S.) **2** (1996), 155-295.

[27] L. Simon, *Lectures on Geometric Measure Theory*, Proceedings of the Centre for Mathematical Analysis **3** (1983), Australian National University.

[28] L. Siebenmann and D. Sullivan, *On complexes that are Lipschitz manifolds*, "Geometric Topology", J. Cantrell, ed., Academic Press, 1979, 503-525.

[29] E. M. Stein, *Singular Integrals and Differentiability Properties of Functions*, Princeton University Press, 1970.

MATHÉMATIQUES – BÂT. 425, CNRS UMR 8628, UNIVERSITÉ DE PARIS-SUD, 91405 ORSAY, FRANCE
E-mail address: guy.david@math.u-psud.fr

DEPARTMENT OF MATHEMATICS, RICE UNIVERSITY, HOUSTON, TEXAS 77251 U.S.A.
E-mail address: semmes@rice.edu

Editorial Information

To be published in the *Memoirs*, a paper must be correct, new, nontrivial, and significant. Further, it must be well written and of interest to a substantial number of mathematicians. Piecemeal results, such as an inconclusive step toward an unproved major theorem or a minor variation on a known result, are in general not acceptable for publication. *Transactions* Editors shall solicit and encourage publication of worthy papers. Papers appearing in *Memoirs* are generally longer than those appearing in *Transactions* with which it shares an editorial committee.

As of November 31, 1999, the backlog for this journal was approximately 4 volumes. This estimate is the result of dividing the number of manuscripts for this journal in the Providence office that have not yet gone to the printer on the above date by the average number of monographs per volume over the previous twelve months, reduced by the number of issues published in four months (the time necessary for preparing an issue for the printer). (There are 6 volumes per year, each containing at least 4 numbers.)

A Copyright Transfer Agreement is required before a paper will be published in this journal. By submitting a paper to this journal, authors certify that the manuscript has not been submitted to nor is it under consideration for publication by another journal, conference proceedings, or similar publication.

Information for Authors and Editors

Memoirs are printed by photo-offset from camera copy fully prepared by the author. This means that the finished book will look exactly like the copy submitted.

The paper must contain a *descriptive title* and an *abstract* that summarizes the article in language suitable for workers in the general field (algebra, analysis, etc.). The *descriptive title* should be short, but informative; useless or vague phrases such as "some remarks about" or "concerning" should be avoided. The *abstract* should be at least one complete sentence, and at most 300 words. Included with the footnotes to the paper, there should be the 1991 *Mathematics Subject Classification* representing the primary and secondary subjects of the article. This may be followed by a list of *key words and phrases* describing the subject matter of the article and taken from it. A list of the numbers may be found in the annual index of *Mathematical Reviews*, published with the December issue starting in 1990, as well as from the electronic service e-MATH [**telnet e-MATH.ams.org** (or **telnet 130.44.1.100**). Login and password are **e-math**]. For journal abbreviations used in bibliographies, see the list of serials in the latest *Mathematical Reviews* annual index. When the manuscript is submitted, authors should supply the editor with electronic addresses if available. These will be printed after the postal address at the end of each article.

Electronically prepared papers. The AMS encourages submission of electronically prepared papers in $\mathcal{A}_{\mathcal{M}}\mathcal{S}$-TEX or $\mathcal{A}_{\mathcal{M}}\mathcal{S}$-LATEX. The Society has prepared author packages for each AMS publication. Author packages include instructions for preparing electronic papers, the *AMS Author Handbook*, samples, and a style file that generates the particular design specifications of that publication series for both $\mathcal{A}_{\mathcal{M}}\mathcal{S}$-TEX and $\mathcal{A}_{\mathcal{M}}\mathcal{S}$-LATEX.

Authors with FTP access may retrieve an author package from the Society's Internet node **e-MATH.ams.org** (130.44.1.100). For those without FTP access, the author package can be obtained free of charge by sending e-mail to pub@ams.org (Internet) or from the Publication Division, American Mathematical Society, P.O. Box 6248, Providence, RI 02940-6248. When requesting an author package, please specify $\mathcal{AMS}$-TEX or $\mathcal{AMS}$-LATEX, Macintosh or IBM (3.5) format, and the publication in which your paper will appear. Please be sure to include your complete mailing address.

Submission of electronic files. At the time of submission, the source file(s) should be sent to the Providence office (this includes any TEX source file, any graphics files, and the DVI or PostScript file).

Before sending the source file, be sure you have proofread your paper carefully. The files you send must be the EXACT files used to generate the proof copy that was accepted for publication. For all publications, authors are required to send a printed copy of their paper, which exactly matches the copy approved for publication, along with any graphics that will appear in the paper.

TEX files may be submitted by email, FTP, or on diskette. The DVI file(s) and PostScript files should be submitted only by FTP or on diskette unless they are encoded properly to submit through e-mail. (DVI files are binary and PostScript files tend to be very large.)

Files sent by electronic mail should be addressed to the Internet address pub-submit@ams.org. The subject line of the message should include the publication code to identify it as a Memoir. TEX source files, DVI files, and PostScript files can be transferred over the Internet by FTP to the Internet node e-math.ams.org (130.44.1.100).

Electronic graphics. Figures may be submitted to the AMS in an electronic format. The AMS recommends that graphics created electronically be saved in Encapsulated PostScript (EPS) format. This includes graphics originated via a graphics application as well as scanned photographs or other computer-generated images.

If the graphics package used does not support EPS output, the graphics file should be saved in one of the standard graphics formats—such as TIFF, PICT, GIF, etc.—rather than in an application-dependent format. Graphics files submitted in an application-dependent format are not likely to be used. No matter what method was used to produce the graphic, it is necessary to provide a paper copy to the AMS.

Authors using graphics packages for the creation of electronic art should also avoid the use of any lines thinner than 0.5 points in width. Many graphics packages allow the user to specify a "hairline" for a very thin line. Hairlines often look acceptable when proofed on a typical laser printer. However, when produced on a high-resolution laser imagesetter, hairlines become nearly invisible and will be lost entirely in the final printing process.

Screens should be set to values between 15% and 85%. Screens which fall outside of this range are too light or too dark to print correctly.

Any inquiries concerning a paper that has been accepted for publication should be sent directly to the Editorial Department, American Mathematical Society, P. O. Box 6248, Providence, RI 02940-6248.

Editors

This journal is designed particularly for long research papers (and groups of cognate papers) in pure and applied mathematics. Papers intended for publication in the *Memoirs* should be addressed to one of the following editors:

Ordinary differential equations, partial differential equations, and applied mathematics to JOHN MALLET-PARET, Division of Applied Mathematics, Brown University, Providence, RI 02912-9000; electronic mail: `jmp@cfm.brown.edu`.

Harmonic analysis, representation theory, and Lie theory to ROBERT J. STANTON, Department of Mathematics, The Ohio State University, 231 West 18th Avenue, Columbus, OH 43210-1174; electronic mail: `stanton@math.ohio-state.edu`.

Ergodic theory and dynamical systems to ROBERT F. WILLIAMS, Department of Mathematics, University of Texas at Austin, Austin, TX 78712-1082; e-mail: `bob@math.utexas.edu`

Real and harmonic analysis and geometric partial differential equations to WILLIAM BECKNER, Department of Mathematics, University of Texas at Austin, Austin, TX 78712-1082; e-mail: `beckner@math.utexas.edu`.

Algebra to CHARLES CURTIS, Department of Mathematics, University of Oregon, Eugene, OR 97403-1222 e-mail: `cwc@darkwing.uoregon.edu`

Algebraic topology and cohomology of groups to STEWART PRIDDY, Department of Mathematics, Northwestern University, 2033 Sheridan Road, Evanston, IL 60208-2730; e-mail: `s_priddy@math.nwu.edu`.

Differential geometry and global analysis to CHUU-LIAN TERNG, Department of Mathematics, Northeastern University, Huntington Avenue, Boston, MA 02115-5096; e-mail: `terng@neu.edu`.

Probability and statistics to RODRIGO BAÑUELOS, Department of Mathematics, Purdue University, West Lafayette, IN 47907-1968; e-mail: `banuelos@math.purdue.edu`.

Combinatorics and Lie theory to PHILIP J. HANLON, Department of Mathematics, University of Michigan, Ann Arbor, MI 48109-1003; e-mail: `hanlon@math.lsa.umich.edu`.

Logic to THEODORE SLAMAN, Department of Mathematics, University of California at Berkeley, Berkeley, CA 94720-3840; e-mail: `slaman@math.berkeley.edu`.

Number theory and arithmetic algebraic geometry to ALICE SILVERBERG, c/o Mathematisches Institut, Universitaet Erlangen–Nuernberg, Bismarckstraße 1 1/2, 91054 Erlangen, Germany; e-mail: `silver@math.ohio-state.edu`.

Complex analysis and complex geometry to DANIEL M. BURNS, Department of Mathematics, University of Michigan, Ann Arbor, MI 48109-1003; e-mail: `dburns@math.lsa.umich.edu`.

Algebraic geometry and commutative algebra to LAWRENCE EIN, Department of Mathematics, University of Illinois, 851 S. Morgan (M/C 249), Chicago, IL 60607-7045; e-mail: `ein@uic.edu`.

Geometric topology, knot theory, hyperbolic geometry, and general topoogy to JOHN LUECKE, Department of Mathematics, University of Texas at Austin, Austin, TX 78712-1082; e-mail: `luecke@math.utexas.edu`.

Partial differential equations and applied mathematics to BARBARA LEE KEYFITZ, Department of Mathematics, University of Houston, 4800 Calhoun, Houston, TX 77204-3476; e-mail: `keyfitz@uh.edu`

Operator algebras and functional analysis to BRUCE E. BLACKADAR, Department of Mathematics, University of Nevada, Reno, NV 89557; e-mail: `bruceb@math.unr.edu`

All other communications to the editors should be addressed to the Managing Editor, PETER SHALEN, Department of Mathematics, University of Illinois, 851 S. Morgan (M/C 249), Chicago, IL 60607-7045; e-mail: `shalen@math.uic.edu`.

Selected Titles in This Series

(Continued from the front of this publication)

655 **Yuval Z. Flicker**, Matching of orbital integrals on $GL(4)$ and $GSp(2)$, 1999

654 **Wancheng Sheng and Tong Zhang**, The Riemann problem for the transportation equations in gas dynamics, 1999

653 **L. C. Evans and W. Gangbo**, Differential equations methods for the Monge-Kantorovich mass transfer problem, 1999

652 **Arne Meurman and Mirko Primc**, Annihilating fields of standard modules of $\mathfrak{sl}(2,\mathbb{C})^{\sim}$ and combinatorial identities, 1999

651 **Lindsay N. Childs, Cornelius Greither, David J. Moss, Jim Sauerberg, and Karl Zimmermann**, Hopf algebras, polynomial formal groups, and Raynaud orders, 1998

650 **Ian M. Musson and Michel Van den Bergh**, Invariants under Tori of rings of differential operators and related topics, 1998

649 **Bernd Stellmacher and Franz Georg Timmesfeld**, Rank 3 amalgams, 1998

648 **Raúl E. Curto and Lawrence A. Fialkow**, Flat extensions of positive moment matrices: Recursively generated relations, 1998

647 **Wenxian Shen and Yingfei Yi**, Almost automorphic and almost periodic dynamics in skew-product semiflows, 1998

646 **Russell Johnson and Mahesh Nerurkar**, Controllability, stabilization, and the regulator problem for random differential systems, 1998

645 **Peter W. Bates, Kening Lu, and Chongchun Zeng**, Existence and persistence of invariant manifolds for semiflows in Banach space, 1998

644 **Michael David Weiner**, Bosonic construction of vertex operator para-algebras from symplectic affine Kac-Moody algebras, 1998

643 **Józef Dodziuk and Jay Jorgenson**, Spectral asymptotics on degenerating hyperbolic 3-manifolds, 1998

642 **Chu Wenchang**, Basic almost-poised hypergeometric series, 1998

641 **W. Bulla, F. Gesztesy, H. Holden, and G. Teschl**, Algebro-geometric quasi-periodic finite-gap solutions of the Toda and Kac-van Moerbeke hierarchies, 1998

640 **Xingde Dai and David R. Larson**, Wandering vectors for unitary systems and orthogonal wavelets, 1998

639 **Joan C. Artés, Robert E. Kooij, and Jaume Llibre**, Structurally stable quadratic vector fields, 1998

638 **Gunnar Fløystad**, Higher initial ideals of homogeneous ideals, 1998

637 **Thomáš Gedeon**, Cyclic feedback systems, 1998

636 **Ching-Chau Yu**, Nonlinear eigenvalues and analytic-hypoellipticity, 1998

635 **Magdy Assem**, On stability and endoscopic transfer of unipotent orbital integrals on p-adic symplectic groups, 1998

634 **Darrin D. Frey**, Conjugacy of Alt_5 and $SL(2,5)$ subgroups of $E_8(\mathbb{C})$, 1998

633 **Dikran Dikranjan and Dmitri Shakhmatov**, Algebraic structure of pseudocompact groups, 1998

632 **Shouchuan Hu and Nikolaos S. Papageorgiou**, Time-dependent subdifferential evolution inclusions and optimal control, 1998

631 **Ronnie Lee, Steven H. Weintraub, and J. William Hoffman**, The Siegel modular variety of degree two and level four/Cohomology of the Siegel modular group of degree two and level four, 1998

630 **Florin Rădulescu**, The Γ-equivariant form of the Berezin quantization of the upper half plane, 1998

For a complete list of titles in this series, visit the AMS Bookstore at **www.ams.org/bookstore/**.